iPad Manual for Beginners

2nd Edition (iOS 11)

Covers iPad Models: Air, Air 2, 5th Generation, Mini 2, Mini 3, Mini 4, & Pro

Joe Malacina

An Infinity Guides book

InfinityGuides.com

Trademarks & Acknowledgements

Warning & Disclaimer

Contact the Publisher

To contact No Limit Enterprises, Inc. or the author for sales, marketing material, or any commercial purpose, please visit www.nolimitcorp.com, or e-mail us at info@nolimitcorp.com.

iPad® Manual for Beginners, 2ⁿᵈ Edition
Publisher: No Limit Enterprises, Inc.
Author: Joe Malacina
ISBN: 978-0-9989196-8-3
Library of Congress Control Number: 2018907588
Printed in the U.S.A.

TABLE OF CONTENTS

INTRODUCTION

Congratulations! So you have decided to take the first step, in fact the only step needed to learn how to use your iPad. Maybe you do not even have an iPad yet, and just want to see how it works before you decide whether to buy one. Either way, this book will teach you everything you need to know on using your iPad. I want to take this time to tell you how this book is going to be the only thing you will ever need in order to learn your iPad. You see, this book was written for the perspective of a beginner. In other words, if you have never used a "tablet" in your life, that will be no detriment when reading this book. That is the big difference between this book and other competitors available. Many authors fail to realize that even in 2018, many people are buying their first "tablet," and need to be shown from the ground up the basics of using and navigating their device. So that is what this book sets out to accomplish. I will teach you not only how to do specific functions on your iPad, but I will teach you the building blocks of using any tablet or "smart" device. When you are finished reading this book, not only will you be a pro at using your iPad, but you will be able to pick up any tablet and have a general understanding of how it works and how to accomplish tasks.

This book is structured so that basic concepts I teach you in beginning chapters will be used in later chapters. Therefore, I highly recommend reading the first few chapters, instead of skipping ahead to exactly what you want to learn. You may miss out on essential tidbits of information that I will not cover in detail in later chapters.

Lastly, you may be wondering if this book is suitable for you. Will it answer all the questions you have? Will you understand the material? I can assure you that the information addressed in this book is derived directly from the input of several thousand iPad users who have had the same questions as you. I have been teaching people how to use their iPhone and iPad for over 5 years, and have taught well over 50,000 people of all ages and backgrounds how to use their devices. I have run a blog and numerous websites where I have received over 10,000 emails with questions on how to do this and how to do that. So I can assure you, I know what the most common questions are, and where the most confusion lies. So take solace in the fact that this book will address your questions with a step-by-step approach, while building your technological intuition. When you are done, you will not even need to memorize the steps to perform a function, you will have the intuition and knowledge to figure it out quickly. That is the core of what this book will teach you.

This being the 2nd edition of the *iPad Manual for Beginners*, every chapter has been updated to better explain and demonstrate the different functions of the iPad. We have also updated many of the illustrations and completely revamped the Recommended Apps Appendix for the year 2018. As always, I welcome your feedback on this book and you can always reach me on Twitter @JoeMalacina.

On a final note, Apple users are notoriously loyal to their electronic devices. After reading this book, you may start to see exactly why. The iPad interface and navigation is seamless and soon you will find yourself liking the way it does things. Soon after that, you will notice that many other Apple devices work in very similar ways, and may find yourself wanting to explore other Apple devices such as an iPhone or Mac. If you ever find yourself wondering how these devices work, I urge you to check out www.infinityguides.com. There you can find beginners' manuals and online courses on many Apple devices.

On that introduction, let us get started

Chapter 1 – About the iPad

So you now have an iPad and you are ready to start using it. So what is a new iPad exactly? Your iPad is classified as a tablet, which means that it is an electronic device that can perform tasks and functions. So with your iPad you can browse the internet, send text messages, check your email, make video calls, and use applications. All of this is done with the iPad and its touch screen. Before I show you how to do all these tasks, let's familiarize ourselves with some key terms that I will use often in this book. These terms are important for you to remember, as you will encounter them often.

Key Terms

Apple ID – Let's get the trickiest one out of the way early. By far, the most common question I receive is about the Apple ID. Let me define exactly what an Apple ID is. If you have ever used an Apple device before, such as a Mac computer or iPhone, then you have probably encountered the term Apple ID before. An Apple ID is just an email address that is associated with you and your Apple devices. You need only one Apple ID, and you can use that same Apple ID on all of your Apple devices. I will go into creating an Apple ID a little later, but for now just know that your Apple ID is your account that you will use on your iPad. This account saves all of your important data such as contacts, photos, and your credit card information.

Apps – Apps are programs on your iPad that can do tasks. Your iPad comes with many apps already installed, and you can download many more through the App Store. Nearly every aspect within the iPad is part of an app, which you will see later on. Apps appear as square icons on your iPad's home screen (See Figure 1.1).

iTunes – iTunes is an app and a piece of software that is used to manage your iPad's data. There is an app on your iPad called iTunes, where you can purchase and download music. More importantly, you can and should download iTunes on one of your computers. iTunes software on a computer backs up your iPad's data every time you plug it in, and iTunes can be used to reset your iPad in case there is a major problem with it. It is not necessary, but highly recommended that you download and install iTunes on one of your computers. Instructions on how to download iTunes can be found on the website www.applevideoguides.com/itunes-install.html.

Home Screen – Throughout this book, you will hear the term home screen used often. Your home screen is the main screen of your iPad where all of your Apps are shown.

Portrait & Landscape – You can view your iPad in portrait or landscape mode. Portrait mode is the standard mode where your home button is on the bottom. Landscape mode is when you turn your iPad horizontal, and the content of your iPad also becomes horizontally oriented. Many apps allow you to use your iPad in landscape mode and some require it. Just know that you can change the orientation of your iPad at any time, and many apps allow you to do so.

*For this book, I will be demonstrating using the iPad in Portrait mode. All illustrations, instructions, and locations are shown while using the iPad in Portrait mode, which is when the circular home button is at the bottom. This is purely for simplicity and to provide the best learning experience.

Different iPads & iOS

There are many different iPads out in the marketplace, and it is important to understand the main differences between them. The reason this is important is so you understand why this book works for nearly all iPads, and here it is: All iPads operate nearly exactly the same, depending upon which software they are using. So if you have an iPad Air 2, and your friend has an iPad Mini 3, they will operate in almost the exact same way in every single aspect as long as you are both using the same software version. The only differences between iPads are where some buttons are located, a few features, and the hardware inside the iPad. Some older iPads will have different charging ports as well. All in all, remember this: No matter which iPad you have, it will work nearly exactly the same as every other iPad that is running the same software. This is even true with the iPad Pro, the most powerful model of the iPad. The main differences between an iPad Pro and other iPads are the size of the screen, the hardware, the ability to use an Apple Pencil, and the special external keyboard you can use with the iPad Pro. Other than that, the iPad Pro will work the same as an iPad Air 2, or any other iPad that is running the same software.

Let's now explain this software running on iPads. The software your iPad is running is called its iOS. iOS stands for I Operating System, and it governs how everything in your iPad works. iOS is labeled by a number, and the higher the number, the newer the software. For instance, the iPad Mini 4 comes with iOS 10. The iPad Mini 3 originally came with iOS 8.1. At all times you should be using the newest version of iOS available for your iPad. So if you have an iPad Mini 3 with iOS 8.1, you should update the software to iOS 11. I will show you how to update the software in Chapter 3.

All of these square icons are apps

The right side of the Dock shows suggested and recently used apps

This bottom section is called the "Dock"

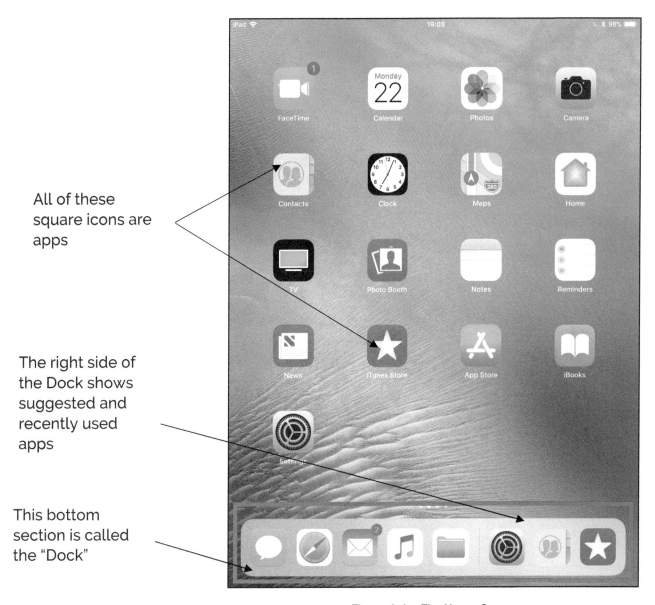

Figure 1.1 – The Home Screen

Chapter 2 – iPad Layout

Now we enter the "instruction manual" portion of this book, and we start with the iPad layout. Shown in <u>Figure 2.1</u> is the layout of an iPad, specifically the iPad Air 2.

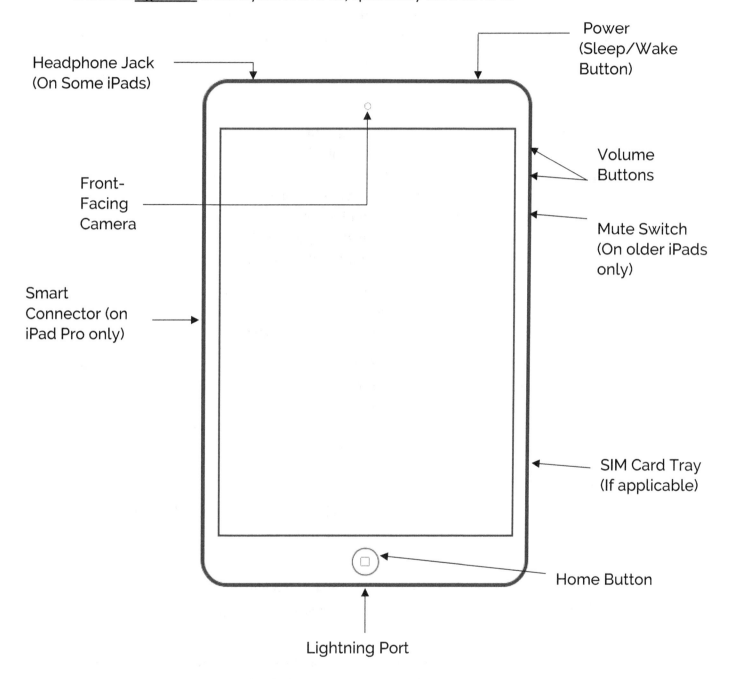

Figure 2.1 – Layout of iPad

As you can see, the power button, also known as the sleep/wake button, is on the top of the iPad on the right side, while the volume buttons are on the right side of the iPad.

Pressing and holding the power button turns the iPad on when the tablet is off. When the iPad is already on, pressing the power button turns the screen on or off. The volume buttons are used to control the volume of the iPad. Pressing the top volume button increases the iPad volume, while pressing the bottom volume button decreases volume. On older iPads there will be a Mute switch below the volume buttons. This switch turns Mute mode on or off. When Mute mode is on, all sounds are turned to silent for alerts. With newer iPads, turning on Mute mode is done through the Control Center, which is covered later in this book.

Looking at the front of the iPad, we have the most important button, the home button. The home button is the circular button on the front of the iPad near the bottom. Pressing this button at any time will ALWAYS return you to your home screen. So if you are inside an app and need to go to your home screen, pressing the home button will accomplish this. The home button is also the button you will use to unlock your iPad from the lock screen. You will also use the home button to scan your fingerprint, if your iPad supports that feature. Also on the front of the iPad near the top, you can see a small circle which is the front facing camera.

At the bottom edge of your iPad is the charging port. On older iPads this will be a wider port than on newer iPads. This port is called the lightning connector port, and allows you to charge your iPad, connect accessories, and connect your iPad to other devices.

If your iPad is cellular capable, you will have a SIM card tray on the right edge of the device. You can open this tray by using a paperclip and gently pushing the end of the paperclip into the small hole of the tray. This will pop open the tray, and you can place your SIM card on it and then push the tray back into the iPad until it clicks into place. A cellular capable iPad can use internet and data outside of a Wi-Fi network and requires a cellular data plan.

On the back of your iPad is the rear camera.

Lastly, if you have an iPad Pro there will be what is a called a "Smart Connector" on the left edge of the device. This is for connecting accessories such as a keyboard.

Turning your iPad on and off

When your iPad is turned off, you can turn it on by pressing and holding the <u>power button</u> until your screen lights up. This will turn your iPad on and it will start booting up. After several seconds it will boot up completely.

At any time you can turn your iPad off by pressing and holding the <u>power button</u> until <u>Figure 2.2</u> appears. To turn your iPad off from this screen, place your finger down on the red power symbol, and slide it across to the right, releasing at the end. This turns your iPad off.

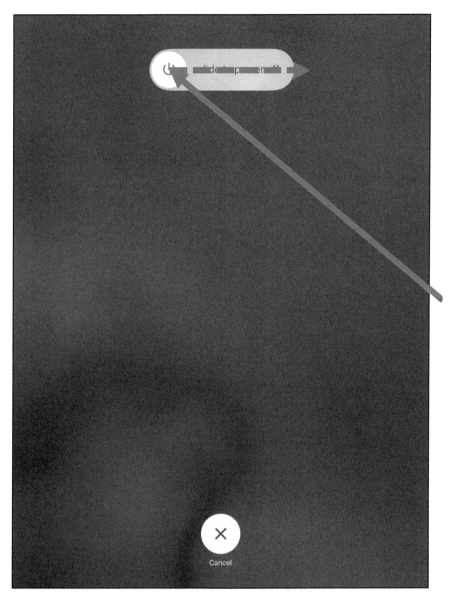

To turn iPad off from this screen: Tap on the power icon with your finger, hold down, then swipe your finger to the right, releasing at the end.

Figure 2.2 – Power iPad off

Charging your iPad

To charge your iPad, plug the included USB cable into the bottom of your iPad, and connect the opposite end (USB) into a USB port or charging dock. Any powered USB port will charge your iPad including USB ports on your computer or vehicle. The charging dock can be plugged into an electrical outlet. Once your iPad is fully charged, you should disconnect the charging cable. For best battery life, the ideal practice you can follow is letting your iPad's battery drain to yellow or red levels, then charge it until it is full, followed by unplugging the charging cable.

Chapter 3 – Getting Started

This chapter covers getting started with your iPad, and goes over the first-time setup procedure. If you have already completed the first-time setup where you chose your language and created your Apple ID then you can skip the first part of this chapter.

First-Time Setup

When you power on your iPad for the very first time (press and hold the <u>power button</u> when the iPad is off), you will be brought to the first-time setup screen shown in <u>Figure 3.1</u>. This first screen will say Hello in multiple languages. Here are the steps required to move past the initial setup successfully. (Please note that these steps may appear in a different order for you depending upon which iPad you have. Regardless, the steps will be similar and you should be able to follow along.)

1. On the Hello screen (<u>Figure 3.1</u>), press the <u>home button</u> one time to move to the next screen. Or, you may need to tap down on the screen with your finger, and swipe your finger to the right to begin.

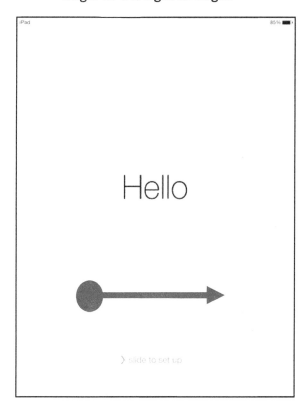

To move to the next screen, tap down on the screen with your finger, hold it there, and slide your finger to the right, and release your finger at the end. This is called a swipe. Alternatively, you can press the home button.

<u>Figure 3.1</u> – Hello screen

2. You are now asked to select your language (<u>Figure 3.2</u>). Tap with your finger one time on the language you wish to use for your iPad. If your language does not appear, tap

down on the screen and move your finger up and down to scroll through the list of languages. Release your finger to complete the scroll.

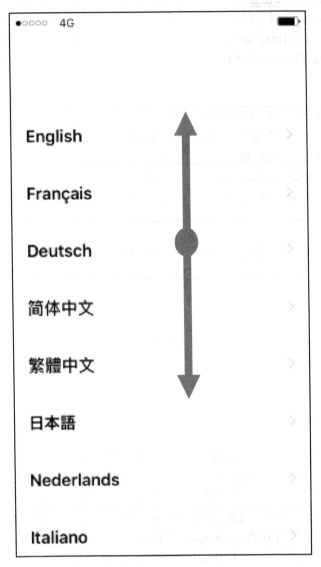

To choose your language, you can tap down on the screen with your finger, and move your finger up and down the screen to "scroll" through different choices. When "scrolling", remove your finger at the end of each movement. Think of it as quickly flicking your finger up or down the screen. To choose your language, tap on the name of the language with your finger and release.

Figure 3.2 – Language screen

3. You will now be asked to select which country or region you live in. Select your country or region by tapping down on the correct one using your finger and release. Again, you can scroll up and down by tapping down and moving your finger up or down and releasing to move throughout the list.

4. The next screen is called the Quick Start screen, and this screen allows you to quickly setup your iPad if you have any other iPhone or iPad around that is running iOS 11 or newer (software version). If you do have one and want to set up your iPad this way, just turn on the other device and bring it next to your iPad. Doing so will set up your iPad with the same settings and Apple ID as your other device. If not, or if you prefer to set

up your iPad manually, tap on the text at the bottom that says <u>Set Up Manually</u>. (In this book, we will be setting up the iPad manually.)

5. The next screen will ask you to choose your Wi-Fi network. It is best to setup your iPad at home so you can connect to your home wireless network. After a few moments, a list of available networks will appear on your screen. Find your home network and tap on it with your finger. A new screen will appear asking for your wireless network password. Use the keyboard that appears at the bottom of your screen to tap in the password. Simply tap each letter or number one at a time to enter it in. You can use the "<u>back arrow</u>" key on the keyboard to backspace, and you can use the "<u>123</u>" key to be brought to numbers and symbols. Also, to capitalize letters, you can use the <u>shift key</u>, which is the key with the upward arrow. Tapping on it will make the letters capitalized for ONE entry. Once you have entered your Wi-Fi password tap <u>Join</u> on your screen (See <u>Figure 3.3</u>), followed by tapping on <u>Next</u> at the upper right.

   a. If you do not have Wi-Fi available near you, or are having trouble connecting to your Wi-Fi network, you can tap on <u>Use Mobile Connection</u> at the bottom of this screen. (Only if your iPad has a cellular data plan)

Tap on the word Cancel to go back and select a different network.

Tap in this box to begin typing your network password.

Tap Join here or at the bottom when finished to join your Wi-Fi network.

This is the "Shift" key i.e. capitalize key.

Backspace key

This is the Symbols & Numbers key. Tap on this to bring up a keyboard with numbers and symbols such as !, ?, etc.

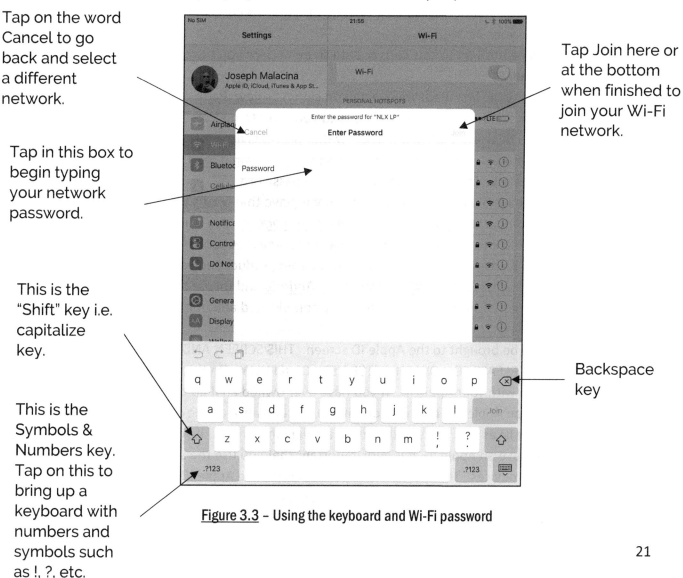

<u>Figure 3.3</u> – Using the keyboard and Wi-Fi password

6. Now your iPad will attempt to activate itself. This can take a few minutes. Once this has completed you will be brought to a new screen to set up Touch ID.

7. On this screen you will be asked to setup Touch ID if your iPad is capable of it. Newer iPads can use Touch ID. For now, tap on <u>Set up Touch ID later</u>, then tap <u>Don't Use</u>. (I will be covering how to set up Touch ID later.)

8. Next, you will be asked to setup a Passcode. A passcode is a 6-digit PIN that allows you to use your iPad. You will need to remember this PIN, so I recommend writing it down somewhere safe. You will have to enter this PIN each time you restart your iPad, and each time Touch ID fails to work properly. I CANNOT STRESS ENOUGH THAT YOU SHOULD WRITE DOWN AND REMEMBER THE PASSCODE YOU CREATE. IF YOU FORGET YOUR PASSCODE, THE PROCESS OF GETTING BACK INTO YOUR IPAD IS DIFFICULT AND TIME CONSUMING. Once you have decided on a passcode, enter it in your iPad by tapping on the numbers. Once you have entered it the first time, you will be asked to confirm the passcode you created by entering it again. Do so and you will be brought to the next screen.
   a. If you prefer, you can use a 4-digit code or alphabetic code instead of a 6-digit passcode. To do this, tap on <u>Passcode options</u>. I will also show you later how to change your passcode.

9. Next you will be brought to the Apps & Data screen. Here you have to choose how to setup your iPad.
   a. If this is your first iPad, you will tap on <u>set up as new iPad</u>. **Most Likely Option.
   b. If you have owned an iPad before, and you want to migrate all your previous data to this new iPad, and you know that previous data is backed up on iCloud, which is Apple's backup service, tap on <u>Restore from iCloud Backup</u>. You will then be asked to enter your Apple ID and password.
   c. If you have owned an iPad before and you have that iPad's data saved on iTunes on a computer, tap on <u>Restore from iTunes Backup</u>. You will then be told to plug your iPad into your computer that has the iTunes backup installed.
   d. Lastly, if your previous tablet was an Android product and you want to migrate that data over, tap on <u>Move Data from Android</u> and follow the instructions on the screen. This option is a little more complicated and not recommended for new users.

10. Next, you will be brought to the Apple ID screen. THIS SCREEN AND THIS STEP IS VERY IMPORTANT. PLEASE FOLLOW THIS STEP VERY CAREFULLY. On this screen you are asked to login with your Apple ID. If you have ever owned an iPad or iPhone, chances are you have an Apple ID. Furthermore, if you currently use a Mac computer, you may have an Apple ID as well. This next sentence is very important. If you have other Apple devices and have an Apple ID, it is very important that you use the same Apple ID for all of your Apple devices. This way, all your purchases on one Apple ID will be available on all of your devices. Just to reiterate from earlier, an Apple ID is just an email address you have appropriately associated with your Apple devices. If you know you already

have an Apple ID, go to step 10A. If you do not have an Apple ID or are unsure, go to step 10B.

    a. If you already have an Apple ID from previous Apple devices, tap into the box to the right of the text "Apple ID" and enter your Apple ID email address. Remember, this is your email address that you have registered as an Apple ID with Apple. After you enter your email address, tap into the box next to password and type in your password using the keyboard that appears. When you are done, tap <u>Next</u> at the upper right. Then you can easily follow the rest of the instructions on your screen.

    b. If you do not have an Apple ID, or this is your first Apple device, or you are not sure if you have an Apple ID, tap on the text at the bottom that says "<u>Don't have an Apple ID or forgot it?</u>"

        i. On the next screen you can tap on "<u>Forgot Apple ID or Password</u>" if you think you have an Apple ID from a previous Apple device. If you do not have an Apple ID, tap on <u>Create a Free Apple ID</u>. **PLEASE NOTE, SETTING UP AN APPLE ID IS HIGHLY RECOMMENDED, AS YOU NEED AN APPLE ID TO DOWNLOAD APPS, BACKUP YOUR DEVICE, AND USE OTHER CRITICAL FEATURES. Some of these steps may appear in a different order for you.

            1. First, you will be prompted to enter your birthday. Do so by tapping in each area at the bottom of your screen and scrolling up or down to select your month, day, and year. Tap <u>Next</u> at the upper right when done.

            2. Now you will be asked to enter your first and last name. Tap into each corresponding box and enter your name. Tap <u>Next</u> at the upper right when done.

            3. Now you will have to enter in some information to create your Apple ID.

                a. Your iPad will now ask which email address you want to use as your Apple ID. You can either use an email address you already have or create a new iCloud email address. I suggest using an email address you already have. Tap <u>Use your current email address</u>.

                b. In the email box, tap into the box next to email and type in the email address you want to use as your Apple ID. I suggest entering your personal email address that you use fairly often. Then tap <u>Next</u> at the upper right.

                c. Next you must create a password for your Apple ID. This box is NOT asking for your email password, it is simply asking you to create a password for your Apple ID. Tap into the password box and type in a password for your

Apple ID. The password must contain at least 8 characters, and it must contain at least 1 capital letter, 1 lower case letter, and at least 1 number.

d. Now tap into the verify box and type in your new password again to verify it is the same. REMEMBER THIS PASSWORD. WRITE IT DOWN SOMEWHERE SAFE IF YOU MUST, BUT YOU WILL NEED THIS PASSWORD LATER. When done, tap Next at the upper right.

e. Next you will need to verify your identity using a phone number or some other contact method. You can enter your cell phone number and then your cell phone will receive a text message or a phone call with a verification code. Get this code and enter it on your screen to verify your identity.

f. Next will be the Terms and Conditions screen. Read these and tap on Agree at the bottom right, if of course you do agree to these terms and conditions.

g. Your Apple ID is now being created. This can take a few minutes, and once it is done you will be brought to a screen that will prompt you to setup your email address that you used for your Apple ID. This step is completely optional. Tap Next at the upper right.

11. The creation of your Apple ID is over. Now you will be brought to an Express Settings page. Tap on Continue at the bottom to proceed. The following steps may appear in a different order for you.

12. On the next screen, you may be asked to setup Apple Pay. Apple Pay lets you pay for items and services using your iPad and is accepted at many terminals. You can choose to set up Apple Pay now by tapping on Continue or you can set it up later on by tapping on Set Up Later in Settings. To set up Apple Pay, you will need a credit card handy.

13. The next screen will be the Siri screen. On this screen you will be asked whether you want to setup Siri. I recommend you set it up now, so tap on Continue. You will be asked to speak a few phrases at your iPad. Follow the directions that appear on your screen. You will have to speak about 5 phrases. All this does is train Siri to recognize your voice. Siri is simply Apple's artificial helper, which listens to you and follows your command. Once this step is complete, your iPad will read "Hey Siri is Ready," and you can tap on Continue. If you are having trouble setting up Siri, just tap on Set Up Siri Later at the bottom of your screen.

14. The next screen will be the iCloud Analytics screen, which is basically asking you whether you want your iPad to periodically send data to Apple to help Apple fix bugs and glitches. This is completely up to you if you want to enable this or not. Tap on Share with Apple or Don't Share to continue.

15. Next a screen with "App Analytics" may appear. This is similar to the previous step. Tap on <u>Share</u> or <u>Don't Share</u>.
16. The next screen will tell you about True Tone Display. Read the information then tap <u>Continue</u>.
17. The next screen will teach you about accessing the Control Center (covered in Chapter 18). Tap <u>Continue</u>.
18. You are all done! You should see the "Welcome to iPad" screen. Press the <u>home button</u> to proceed and to access your home screen.

After you have completed the initial setup you will be brought to your home screen. Remember, your home screen is the screen that shows all of your apps.

Chapter 3.5 – Update your iOS Now

Before we begin Chapter 4, it is important that you check to see if an update is available to your iPad's software, otherwise known as iOS. In order to do this, follow these steps:

1. On your home screen, which shows all of your apps, there should be an app called Settings. Tap on the app with your finger to open it. (Figure 3.5.1)

Figure 3.5.1 – Settings app on Home Screen

2. Next, tap on the word General, which you should see somewhere on the left side of your screen. You can scroll up and down on the left if you cannot find it.

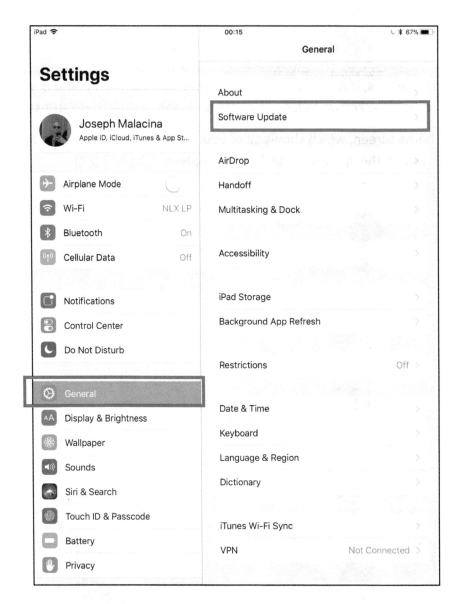

Figure 3.5.2 – Settings -> General -> Software Update

3. Next, tap on <u>Software Update</u>.
4. Your iPad will check to see if there is an update available. If there is an update available, your iPad will say so, and there will be an option to <u>Download/Install</u>. If this option appears, tap on <u>Download and Install</u> or <u>Install</u>. If there is not an update available, your iPad will say iOS X.y.z, where the X, y, and z will be numbers. For the sake of this book, the only number that matters is the first number. So for this book, you will want that first number to be 11 or greater. If that number is an 8, 9, or 10, that is okay as well, but you should still update to the latest version possible. It will also say your Software is Up to Date.

5. If an update is available, install the update by tapping on <u>Install</u> or <u>Download and Install</u>. Your iPad will download and install the software update. This could take some time, and you will need to be connected to Wi-Fi to download the update and have at least 50% battery. (See Chapter 4 to learn how to connect to Wi-Fi if you have not done so in the initial setup). Let the update install, and your iPad will restart when it is finished, and then you are ready to continue. If your iOS software is already up to date, you are ready to continue.

6. You can press the <u>home button</u> on your iPad to return to the home screen.

If a software update is available for your iPad, it is HIGHLY recommended that you update it.

Chapter 4 – Navigating your iPad

All you need to use your iPad are your fingers. Everything is based on the touch screen and the home button on your iPad. Here at the home screen (Figure 4.1), which is just the screen that shows you all of your apps, you can open an app, which is a program on your tablet that can do tasks and enable features, by lightly touching down on the square with your finger and releasing quickly. To then leave the app and return to the home screen, press down on the home button.

Your home screen is actually made up of multiple screens, to accommodate however many apps you have on your iPad. To move through different pages of apps on your iPad's home screen, touch down on the screen with one of your fingers, and then drag your finger to the left or the right, like you would be dragging a page, then release. If at any time you need to return back to the main home screen, just press the home button.

Tap on an App to open it.

Tap and swipe down on your home screen to bring up Spotlight Search

Tap and swipe your finger left and right to switch between different home screens

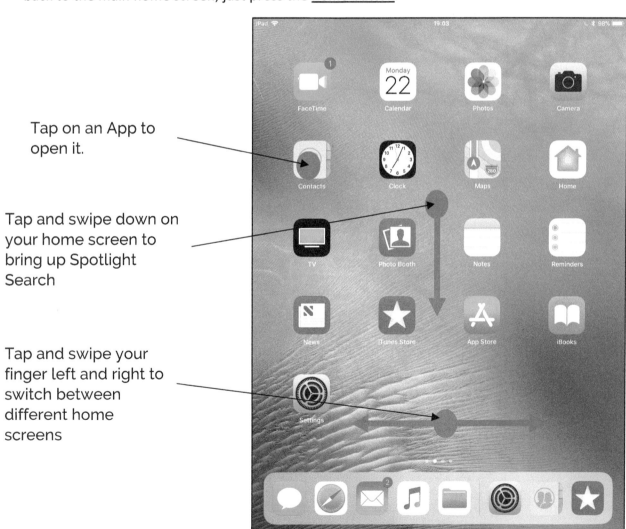

Figure 4.1 – The Home Screen

Spotlight Search

You can search your iPad for anything using the Spotlight Search function. To use this function, touch down somewhere on your home screen, and drag your finger down then release (Figure 4.1). This is a great way to search for apps on your iPad when you have too many to remember them all. To get back to the home screen simply press the home button again, or swipe up with your finger.

Today Screen

You may have noticed that when you are on your main home screen and you swipe your finger to the right that you are brought to what is called the "Today" screen (Figure 4.2). This screen shows you if you have anything coming up on your calendar, along with some suggestions for apps that you use. There may also be more information on this screen such as your local weather. You can tap on anything here to explore it more. To get back to the home screen press the home button or swipe to the left with your finger.

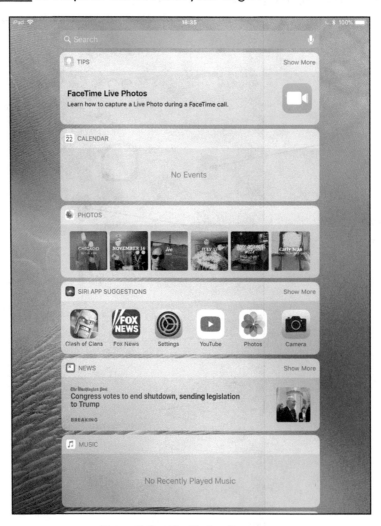

Figure 4.2 – The Today Screen

Connecting to Wi-Fi

Connecting to Wi-Fi is one of the most important things you should do with your iPad. When you are connected to Wi-Fi, your internet speeds are generally faster, AND you are not using valuable cellular data, if applicable. You may have connected to your home wireless network when you first set up your iPad, in which case you are already good to go. You can tell if you are connected to Wi-Fi by looking for the Wi-Fi symbol at the top left of your screen. If the Wi-Fi symbol appears, it means you are connected to Wi-Fi (Figure 4.3). Once you have connected to a Wi-Fi network, your iPad will remember that network, and you will not need to ever manually connect to it again. Each time your iPad comes in range of that network it will connect automatically.

To connect to a Wi-Fi network, follow these steps:

1. Open the Settings app on your home screen by tapping down on it and quickly releasing.
2. Tap Wi-Fi on the left.
3. Your iPad will search for networks. If you are already connected to a network, it will appear directly under the words Wi-Fi and a checkmark will be next to its name. Look at the network names under where it says "CHOOSE A NETWORK..." When you see the network you want to connect to, tap on it with your finger.
4. You will then have to enter your Wi-Fi password. Enter this password using the keyboard, and then tap on Join at the upper right. (For more on using the iPad keyboard, see Figure 3.3 in Chapter 3).
5. Your iPad will join the Wi-Fi network as long as the password was entered correctly. Remember, Wi-Fi passwords are case sensitive. Once you have joined a Wi-Fi network, you will never have to manually join that network again, as your iPad will connect to it automatically from now on.
6. Press the home button to return to the home screen.

TIP: If you do not know your home wireless network name and password, it may be located on the back or bottom of your wireless router. Check the back of your router to see if that information is there.

When you see this Wi-Fi symbol, it means you are currently connected to a Wi-Fi network.

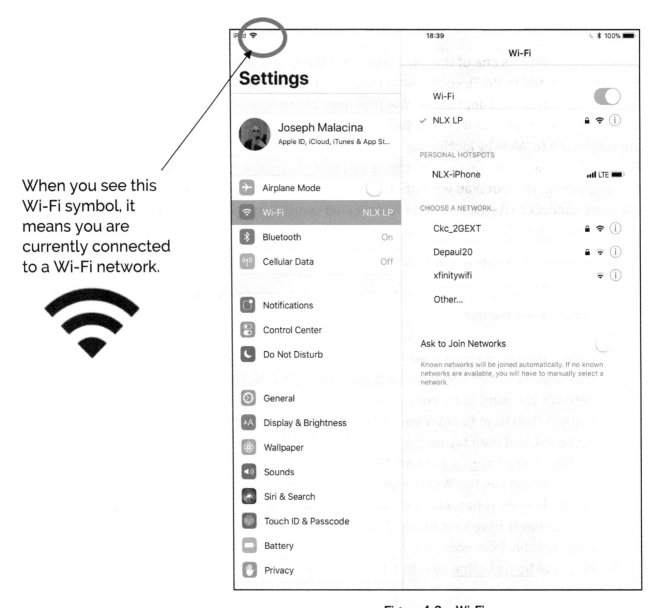

Figure 4.3 – Wi-Fi

Chapter 5 – Apple ID

I have already discussed the concept of your Apple ID throughout this text, but it is worth emphasizing some more. The most common questions I receive about using iPads are related to Apple IDs. So in this chapter, I will cover everything you need to know about your Apple ID.

How to Check Your Apple ID

If you created an Apple ID during the initial setup of your iPad, great! If you did so, please write down your Apple ID email and password somewhere safe. You will need that information from time to time. If you have not created an Apple ID or are not sure, there is a way to check. Follow these steps:

1. Open the <u>Settings app</u> on your iPad by tapping on <u>Settings</u> on your home screen.
2. Scroll down through the Settings options on the left until you find <u>iTunes & App Store</u> and then tap on it.
3. On this screen, at the top, will be a box labeled Apple ID: and inside this box will be your Apple ID if you have created one and are signed in. There should be an email address in there. If there is not an email address in there, it will read "Sign In" instead. (See <u>Figure 5.1</u>)

Now you know whether you are signed in to your iPad with an Apple ID, and what that Apple ID is. If you are signed in, write down and remember your Apple ID. If you are not signed in, and do not have an Apple ID, you need to create one now.

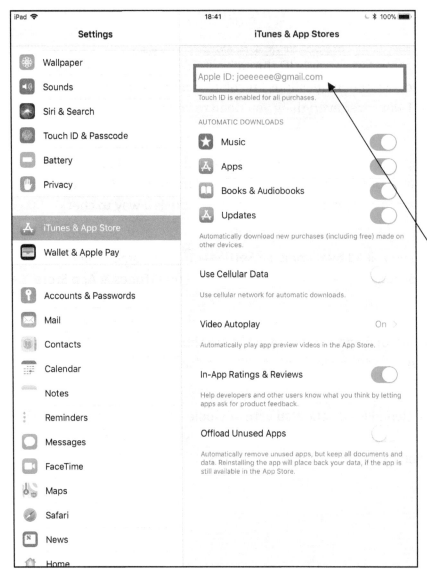

This box here will show your Apple ID. If you are not signed in with your Apple ID, it will say Sign In, which you can tap on.

Figure 5.1 – Settings -> iTunes & App Store

How to Create an Apple ID

If you do not have an Apple ID, you need to create one. Simply put, you need an Apple ID to use many of the functions available on your iPad. To create an Apple ID, follow these steps:

1. Go to your home screen. (Remember, you can get to your home screen at any time by pressing the home button once).

2. Find the app called iTunes Store and tap on it to open it.
3. When the app loads completely, scroll down to the very bottom and tap Sign In.
4. If you have an Apple ID, sign in with your Apple ID login credentials by tapping on Use Existing Apple ID. If you do not have one, tap on Create New Apple ID.

5. On the next screen, enter an existing email address that you use regularly. This email address will become your Apple ID.

6. Next you will have to create a password for your Apple ID. This is not the same as your email's password. It is a separate password for your Apple ID. Type in a password and type it in again in the Verify box. Passwords must contain at least 8 characters, and must contain at least one number, one capitalized letter, and one lowercase letter.

7. Choose your region if it is not automatically filled in for you and agree to the terms and conditions (if applicable) by reading them and tapping on the tab next to them.

8. When done, tap <u>Next</u> at the upper right.

9. At the next screen you will have to fill in your personal information, and select security questions for your account. The security questions are for in case you forget your password. Fill everything in, along with the security questions and answers, and tap <u>Next</u> when done.

10. Next you will have to enter billing information. You will not be charged anything, but this billing information is required in case you decide to purchase anything in the future and to verify your identity. Fill out the form and then tap <u>Next</u> at the upper right.

11. There are a few more steps left that are pretty self-explanatory. Follow the instructions on your screen to complete the creation of your Apple ID. You may be asked to verify your phone number and/or your email address.

12. When all the steps have completed, your Apple ID will have been successfully created. Occasionally, you may be asked for your Apple ID password on your iPad. When you are asked, enter it in on your keyboard. Remember to use correct capitalization.

*Alternatively, you can create an Apple ID from a computer by visiting https://appleid.apple.com/account (Recommended if creating an Apple ID on your iPad becomes too tedious or confusing).

Signing in with Your Apple ID

If you have created your Apple ID on your iPad, you will most likely be signed in automatically. Your Apple ID is used for many apps, and it may be necessary to sign in to all of these apps with your Apple ID. Once signed in, you will not need to ever do this again. Follow these steps to sign in to your iPad with your Apple ID:

1. On your home screen open the <u>Settings app</u>.
2. Scroll down to <u>iTunes & App Store</u> and tap it.
3. Tap the Apple ID box at the top and tap <u>Sign In</u>. If your Apple ID already appears, you are already signed in.
4. Enter in your Apple ID email address and password and sign in.
5. Now scroll on the left and tap <u>on the big box at the very top where it says Apple ID or your name</u> (<u>Figure 5.2</u>).

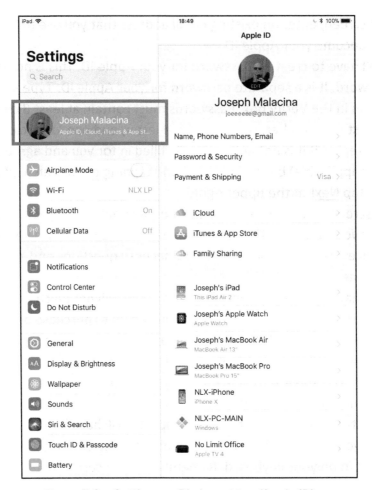

Figure 5.2 – Settings -> Big box at top (Apple ID)

6. At the top will be your name and email address if you are already signed in to your Apple ID for iCloud. If your name does not appear, tap on Sign In.

7. Enter your Apple ID email address and password and tap Sign In.

8. Now we want to check and make sure your Apple ID is signed in for all Apple Services.

9. Find in Settings the Messages box on the left and tap it.

10. On the screen at the right, if there is no box for you to enter your Apple ID email and password, then you are already signed into Messages services with your Apple ID. If this is the case, your screen will say iMessage at the top right. On the other hand, if there is a box to enter your Apple ID email and password, then tap into those boxes and enter the corresponding information and then tap Sign In. (Figure 5.3)

11. Now scroll through the options on the left and tap on FaceTime.

12. In the Apple ID box, make sure your Apple ID appears, if it does not appear, tap on the box and sign in with your Apple ID.

13. Press the home button to return the home screen.

You should now be signed in completely with your Apple ID.

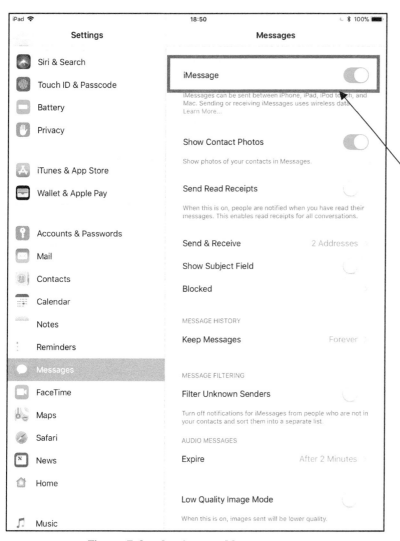

If you see this box with the green switch tab then you are already signed into iMessage with your Apple ID, move on to the next step. Otherwise, enter your Apple ID and password.

<u>Figure 5.3</u> – Settings -> Messages

How to Use Your Apple ID

Now that you are all signed in with your Apple ID, how can you use it? You do not really need to worry about that actually. You will be using your Apple ID automatically. Every time you download an app you will be using your Apple ID. Occasionally, your iPad may show a popup box asking you for your Apple ID password. When this appears, just enter your password and tap <u>OK/Sign In</u>. Your iPad will ask for its password periodically for security purposes. Make sure you remember this password!

All the confusing stuff is now out of the way. Once you are signed in with your Apple ID, you do not need to worry about it anymore as long as you remember your Apple ID password. It is finally time to get into actually using the iPad.

Chapter 6 – Your Contact List

Your contact list is an essential tool on your iPad. With it, you can store your contacts' phone numbers, email addresses, and other vital data.

Importing Contacts

When you first get your new iPad, there will be no contacts in your contact list. If you previously have an iPhone or other Apple device, you can automatically transfer those existing contacts into your iPad. All you need to do to accomplish this is make sure you sign in with the same Apple ID on both devices. Once you are signed in with the same Apple ID on both devices, your devices will share these contacts and they will automatically populate.

There are other ways of transferring contacts into your iPad. For instance, you can transfer your Microsoft Outlook contacts from your computer into your iPad using software known as iTunes. This is much more complicated, and can be learned by visiting www.infinityguides.com and clicking on Online Courses at the top, followed by selecting *How to Use iTunes for Beginners*.

The final option is to create your contacts manually, which is demonstrated next.

Creating Contacts

Let's see how we can create a contact. Here are the steps: (See Figures 6.1 & 6.2)

1. Open the Contacts app from your home screen.
2. At the upper left of your screen, there will be a + sign. Tap on the + sign. (Figure 6.1)
3. You will now be brought to a screen where you can create a contact.
4. Tap into each corresponding box to type in an entry for the new contact. You can enter in their First name, Last name, and Company.
5. Tap into the add phone box to add their phone number.
6. Tap into the add email box to add their email, if applicable.
7. By scrolling down further, you can choose to enter in additional information including their home address, nickname, and other information.
8. You can also choose a specific ringtone for a contact.
9. Once you have added all the information you want for your new contact, tap Done at the upper right.
10. Your new contact has just been created.

Tap to create a new contact

Tap on a contact name to bring up their information on the right

Tap on any of the letters here to be brought to all contacts whose last name starts with that letter

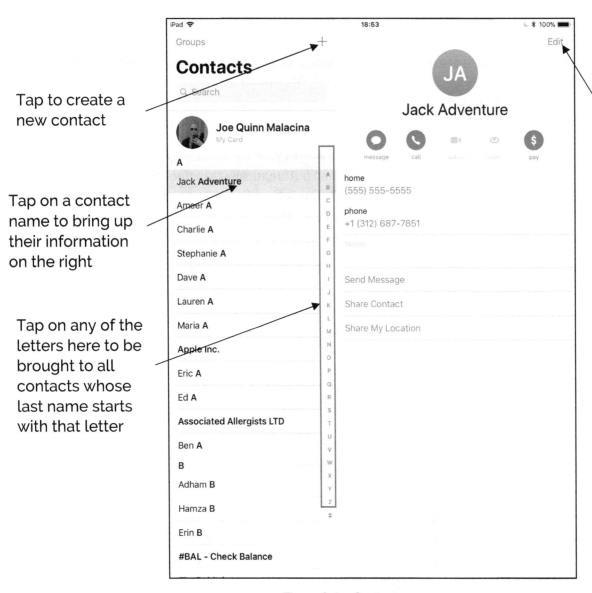

Tap edit to edit a contact

Figure 6.1 – Contacts app

When done tap here.

Tapping on <u>add photo</u> will allow you to set a picture for your contact that is saved in your Photos.

You can tap into each box to type in contact information.

Tap the green <u>+</u> icon to add the corresponding information such as a phone number or email.

You can scroll (tap down on screen and swipe finger up/down) to enter more information for your contact.

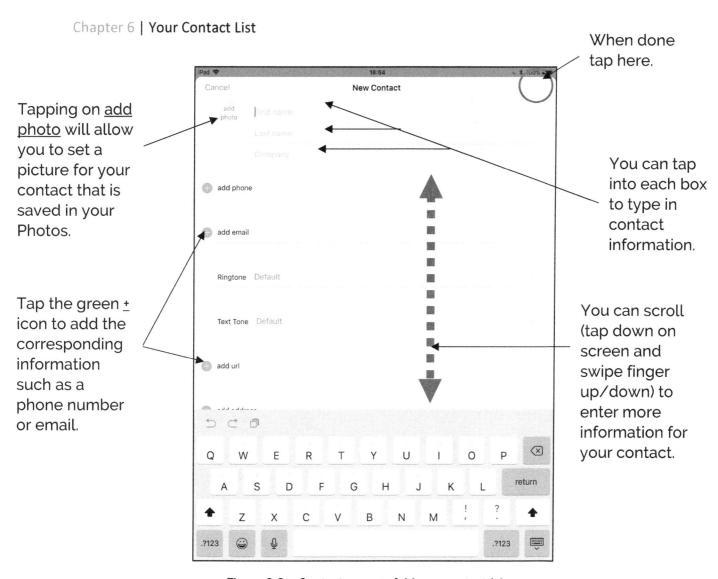

Figure 6.2 – Contacts app -> Add new contact (+)

Browsing Contacts

At any time, you can view your contact list by opening the <u>Contacts</u> app. Now you will see a list of all of your contacts. You can browse through this list by simply tapping down on the left side of your <u>screen</u> and scrolling up or down by swiping your finger up and down on the screen.

TIP: To quickly browse through your contacts by their last name. Tap down and scroll on the <u>letters</u> on the right side of your contact list.

You can also tap into the <u>search bar</u> at the top of your contacts list to search for a contact directly.

To view a contact's information, simply tap on their *name* and their information will appear on the right.

Editing or Deleting a Contact

To edit a contact, tap on their _name_ inside the contact list, and then tap Edit at the upper right. Now you can edit all of their information. To delete a certain aspect of their information, simply tap on the red minus sign next to it. When you are done editing a contact tap on Done at the upper right.

To delete a contact, tap on their _name_ inside the contact list. Now tap Edit at the upper right. Now scroll down all the way to the bottom of the contact's information, and tap on Delete Contact. A confirmation will appear, where you can tap on Delete Contact again to confirm the deletion.

Chapter 7 – Email

Your iPad has the ability to seamlessly manage your email accounts. It is quite convenient when you can quickly reply to an email in 10 seconds straight from your iPad. All of your emails and email accounts are located within the Mail app, which we will get to shortly. For now, let's start by adding your existing email account to your iPad.

Adding an Email Address to iPad

1. Open the Settings app
2. Find the Accounts & Passwords tab on the left, and tap on it
3. Tap Add Account
4. Look through the list of pre-populated email accounts such as AOL, Google (Gmail), and Yahoo. If you have one of these accounts, tap on it. If your account does not appear, tap on Other. (Figure 7.1)

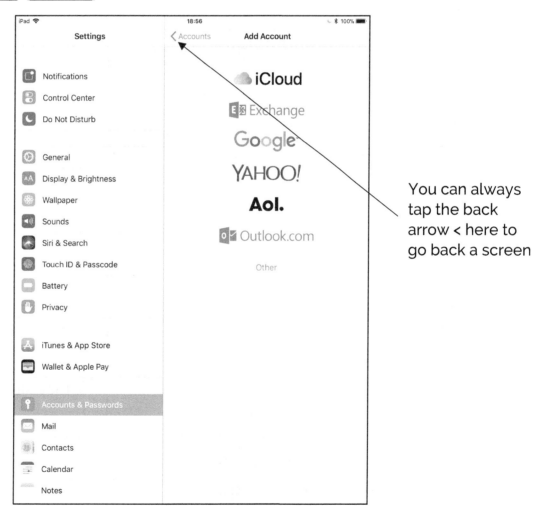

You can always tap the back arrow < here to go back a screen

Figure 7.1 – Settings -> Accounts & Passwords -> Add Account

5. If you tapped on a pre-populated account, follow the instructions that appear on your screen. Most likely you will just have to enter in your email address and password. If you tapped on Other, tap on <u>Add Mail Account</u>.

6. Now type in the requested information. When done, tap <u>Next</u> at the upper right. (<u>Figure 7.2</u>)

7. Your iPad will attempt to add your email account given the information you provided. For most email accounts, this should work with no issues. However, for less common email accounts, your iPad may acquire more information such as incoming and outgoing server information. If this is the case for you, you will have to contact your email provider in order to get this information.

8. Continue following the instructions that appear on your screen until you have reached the end. You can continue by tapping on <u>Next/Done/Save</u> at the upper right.

Once your email account has successfully been added, you are ready to use email on your iPad. You can add multiple email addresses to your iPad if you wish.

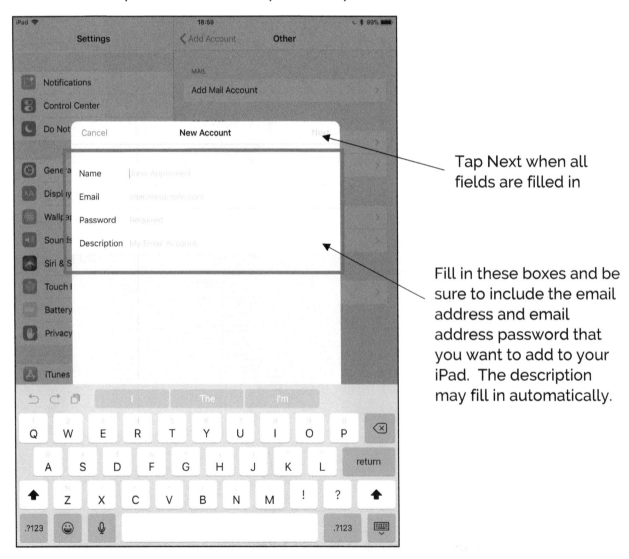

Tap Next when all fields are filled in

Fill in these boxes and be sure to include the email address and email address password that you want to add to your iPad. The description may fill in automatically.

<u>Figure 7.2</u> – Enter your email address information here

Checking Your Email

To check your email, we need to use the Mail app, which is located somewhere on your home

screen. Tap on it to open it up.

Use the back arrow at the upper left to view all of your email accounts (Figure 7.3).

Tap to go back and see all of your mail accounts and mail folders

Manage e-mail in bulk

The blue dot icon means unread mail

Tap on any email message to read it

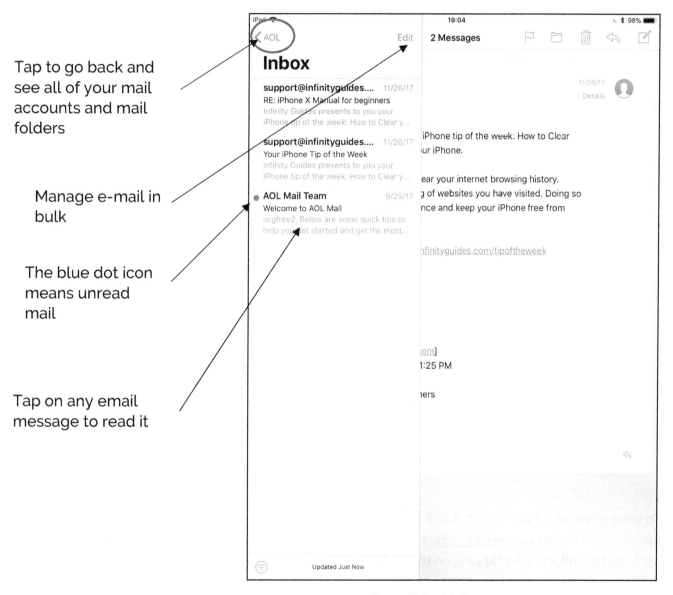

Figure 7.3 – Mail app

Simply tap on an account to view your emails for that specific account. (<u>Figure 7.4</u>)

Here you can see all of your mail accounts and mail folders. Tap on any one of these to explore them.

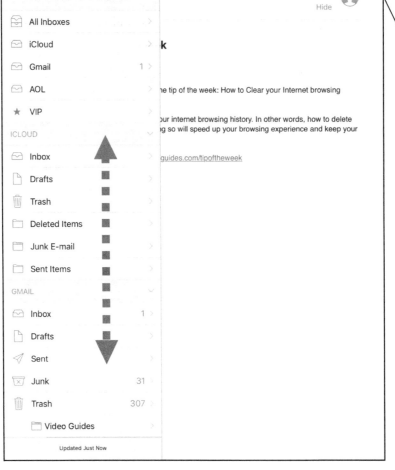

<u>Figure 7.4</u> – Accounts view (Mail app)

Viewing Email

To view an email, simply tap on it and it will appear full screen. From here you have several options available to you (<u>Figure 7.5</u>). Looking at the icons at the top of your screen, they perform the following by tapping on them (Email controls).

- **Flag** – Flags or marks the message.
- **Folder** – Moves the message to a different folder.
- **Trash** – Deletes or archives the message.
- **Arrow** – Replies, forwards, or prints the email.
- **Square with Pencil** – Creates a new email.

You can use the <u>back arrow</u> at the upper left to go back.

Tap to
go back

Tap to
quickly
view next
or
previous
email

Email controls

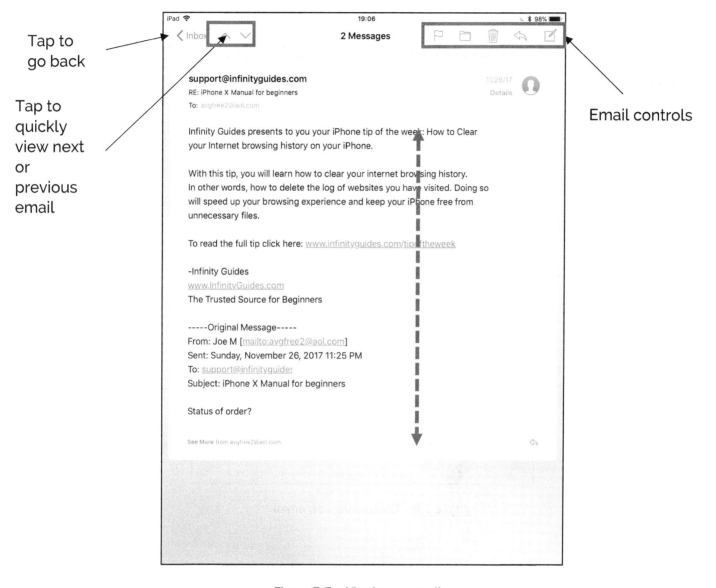

Figure 7.5 – Viewing an email

Sending Email
(See Figure 7.6)

At any time from Mail app, you can create a new email by tapping the square and pencil icon at the upper right. This will bring up a New Message box. Full steps are:

1. Tap the square and pencil icon at the upper right inside the Mail app.
2. Enter in the email address or contact name of the intended recipient. Use the + icon to search through your contacts.
3. Tap into the subject line and type in the subject.
4. Tap into the message body and type in the email message.
5. Tap Send at the upper right, or tap Cancel at the upper left to cancel the message.

Enter recipient email address or contact

Enter email subject

Enter email message

Tap to send

Tap to search your contact list

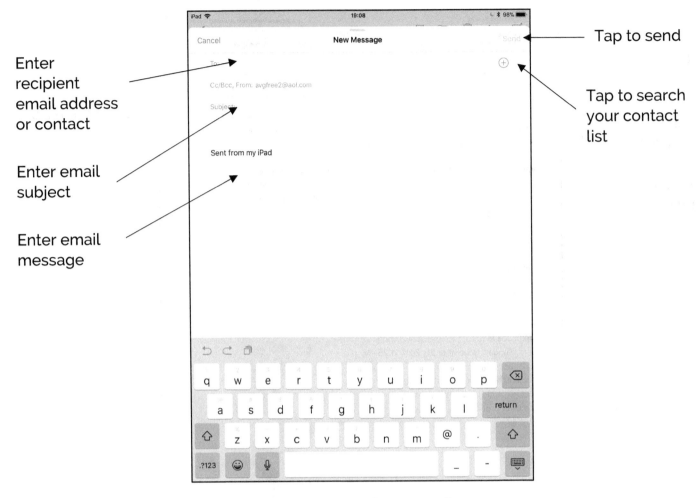

Figure 7.6 – Composing new email

Managing Email

Remember, use the back arrows at the upper left to access your email accounts and folders. If the back arrow does not appear at the upper left, tap at the very upper left with your finger and it will appear.

To Delete Email(s)

1. Tap an email account or folder to view all emails in that account or folder.
2. Tap Edit at the upper right of your email list
3. Tap each email you want to manage
4. Use the text at the bottom to manage. Tap on Mark, Move, or Archive/Delete.
5. To cancel, tap Cancel at the top.

TIP: You can quickly delete an individual email by tapping on its preview, then swiping to the left.

Chapter 8 – Web Browsing

One of the signature features of the iPad is the ease in which you can surf the web. To access the Internet, open the Safari app on your home screen. Safari is where you will be doing all of

your internet browsing, unless you download a different app for web browsing.

Visiting Web Pages

Safari is Apple's web browser optimized for the iPad. In Safari, you can visit web pages as well as search the web using your favorite search engine. By default, Google is set as the current search engine. To search the web, tap in the box at the top of the screen and type in your search parameters, then touch GO on your keyboard. If you want to visit a webpage directly, touch in the same box and type in the web address, followed by tapping GO. (Figure 8.1)

Web page controls

Manage tabs

Tap into the bar to search the web or go directly to a web address.

Manage webpage

Your frequently visited websites and bookmarks will appear here by default. You can tap on the page to go there.

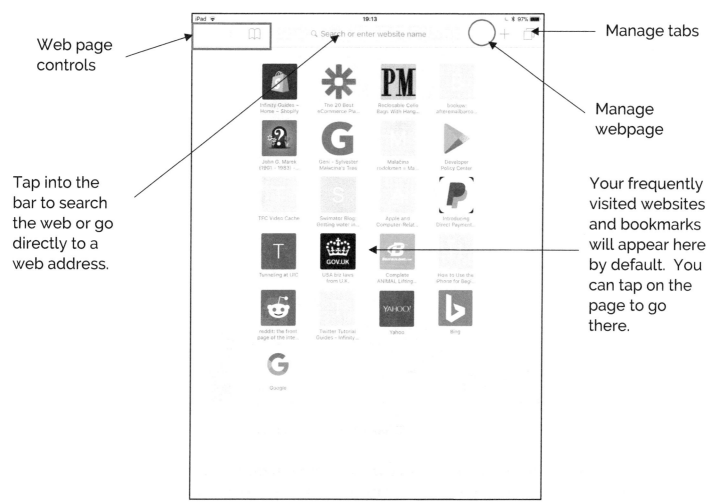

Figure 8.1 – Safari: search web & visit websites

51

Navigating the web on the iPad is a unique experience and takes some practice. Think of your finger as the mouse pointer, and touching down on the screen as a click. To open a link, simply touch down on it with your finger. (Figure 8.2)

Tap on the page to perform a classical *click of the mouse*. Do this to open links and pages.

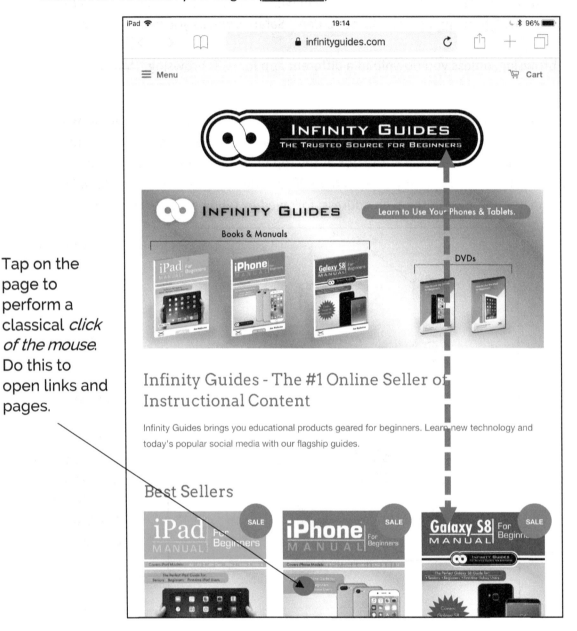

Figure 8.2 – Safari: tap to click (open pages)

Scrolling through web pages works the same way as scrolling through apps. Simply drag your finger up and down to scroll through a web page. If you need to zoom in on something, there are 2 different ways. To zoom in on a specific paragraph or section of a web page, double tap with your finger on that section, and Safari should zoom in and center the screen automatically. To zoom in on a very specific spot of a web page, touch down on the screen with two fingers, with your two fingers being very close together. Then spread your fingers apart while remaining on the screen. To zoom back out, touch two fingers down on the screen again, only this time have the fingers start far apart, and pull the fingers in towards each other. (Figures 8.3 & 8.4)

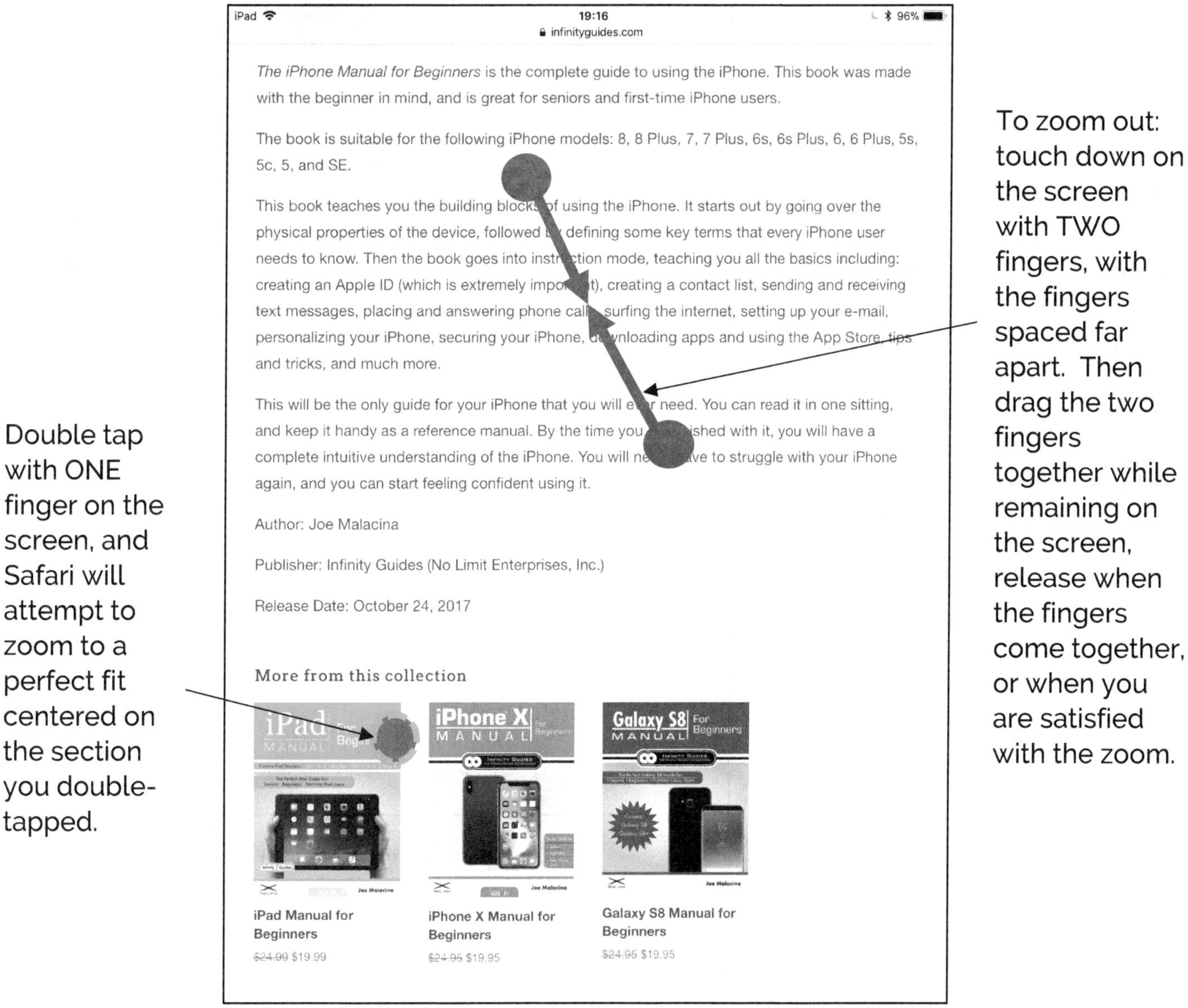

Double tap with ONE finger on the screen, and Safari will attempt to zoom to a perfect fit centered on the section you double-tapped.

To zoom out: touch down on the screen with TWO fingers, with the fingers spaced far apart. Then drag the two fingers together while remaining on the screen, release when the fingers come together, or when you are satisfied with the zoom.

Figure 8.3 – Safari: zooming out

To zoom in: touch down on the screen with TWO fingers, with the two fingers being close together. Then drag the fingers away from each other in opposite directions while remaining on the screen, releasing at the end or when you are satisfied with the zoom.

Figure 8.4 – Safari: zooming in

You can still do all the basic functions of your normal web browser with Safari. To browse through basic web functions, tap down at the top of your screen. These icons are your navigation options also known as web page controls (See Figure 8.1). From here, you can browse as you please. For instance, to go back to the previous page, touch the left arrow. To go forward a page, touch the right arrow. To manage a webpage, touch the rectangle with the arrow in it. From here you have the options to mail the webpage to a friend, message the page to a friend, tweet the page on your Twitter, share the page to Facebook, add a shortcut to the web page to your home screen, print the web page, bookmark the page, or to add the page to your reading list. (Figure 8.5) If you choose to bookmark the page, the page will appear in your bookmarks screen which can be accessed by touching the book icon at the top of Safari.

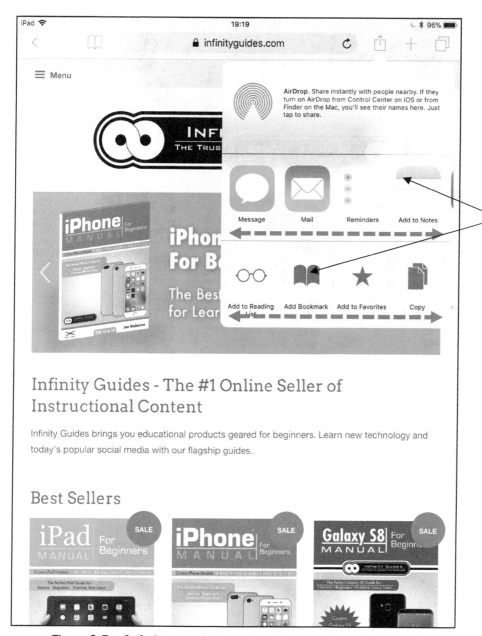

Figure 8.5 – Safari: managing a web page (arrow & rectangle icon)

Finally, you can browse in multiple tabs in Safari, by using the <u>+</u> icon and the <u>two windows icon</u> at the top right of the Safari screen (<u>Figure 8.6</u>). Tabs are just additional browsing windows where you can browse the internet; an excellent tool for multi-taskers. To browse between tabs, touch the <u>windows icon</u> again. To truly master browsing through Safari, just take some time practicing with it, and you will get it in no time.

<u>Windows Icon</u> – Allows you to view and manage all the tabs you have open.

<u>+ Icon</u> – Creates a new browsing tab in Safari.

Tap the x or swipe the page off the screen to close a tab.

Exits tab manager and returns to Safari main screen

Creates a new tab

Switches you to private browsing (does not save cookies or history)

Figure 8.6 – Safari: using Tabs (Windows Icon-see Figure 8.1)

Chapter 9 – Using your Camera

Your iPad is equipped with a powerful camera. To access the camera, touch the Camera app on your home screen. [camera icon] There are a few options available inside the Camera app. On the right of your screen from top to bottom is: (Figure 9.1)

- **[A] - HDR** – This enables or disables High Dynamic Range photography
- **Concentric Circles (if equipped)** – This enables or disables LIVE photos, which capture a second or so before and after the photo, thus creating what is called a "live photo."
- **[B] - Timer** – Tapping on this icon allows you to set a timer for the capture sequence.
- **[C] - Lens Selector** – Allows you to switch between the outward lens and the selfie lens which faces you.
- **[D] - Capture Icon** – Performs action such as capture photo or record video
- **[E] - Photo Preview** – Takes you to your recent photos

Taking a Photo

To take a photo, aim your camera and then tap on the large circle icon at the right of your screen. You can also turn your iPad to the side to take a picture in landscape mode. Furthermore, you can use the selfie lens, which faces directly back at you, by tapping on the camera icon, which is just above the capture icon. (Figure 9.1)

Camera Options

You can switch between different camera modes by tapping down somewhere on your screen while in the Camera app, and then swiping up or down. You will notice different text become highlighted right below the photo preview to indicate which mode you are in. (Figure 9.1)

The different modes are:

- **Photo** – Standard photo mode
- **Video** – Records a video
- **Slo-Mo** – Records a video in slow motion
- **Time-Lapse** – Records a time lapse sequence
- **Square** – Captures a photo in a perfect square
- **Pano** – Captures a panoramic photo.

You can preview the photo you have just taken by tapping on the preview of the photo directly underneath the capture icon. To get back to the camera from this preview, tap Camera at the upper left.

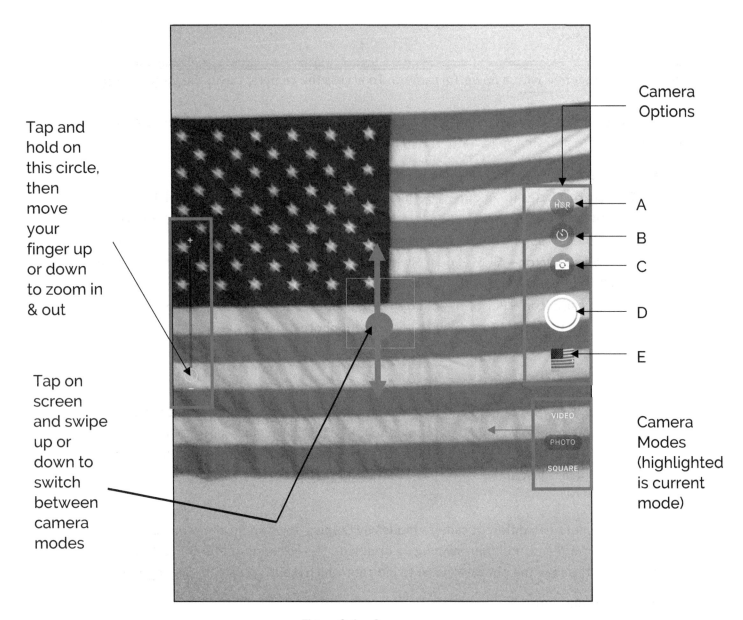

Tap and hold on this circle, then move your finger up or down to zoom in & out

Camera Options

A

B

C

D

E

Tap on screen and swipe up or down to switch between camera modes

Camera Modes (highlighted is current mode)

Figure 9.1 – Camera app

Chapter 10 – Photos & Videos

Now that we know how to take photos and record videos, it is time to learn what we can do with these. All of your photos and videos can be found in the <u>Photos app</u>, which is located

somewhere on your home screen.

Photos App Layout

See (<u>Figure 10.1</u>)

When you first open the <u>Photos app</u>, it may look like there is a lot going on. Let's take a look at navigating this expansive app. At the bottom will be your browsing tabs. The <u>Photos</u> tab will show you all of your photos sorted by date. You can browse through these by scrolling up and down. <u>Memories</u> will show a specific memory, which are automatically generated by your device. An example of a memory is a photo you took one year ago from today. The <u>Shared</u> tab will bring you to shared photo streams, which I will discuss a little later on. Lastly, the <u>Albums</u> tab will bring you to all of your albums. At any time you can view a photo in full screen by simply tapping on it.

Create a
new
album

Manage
albums
(deletion,
etc.)

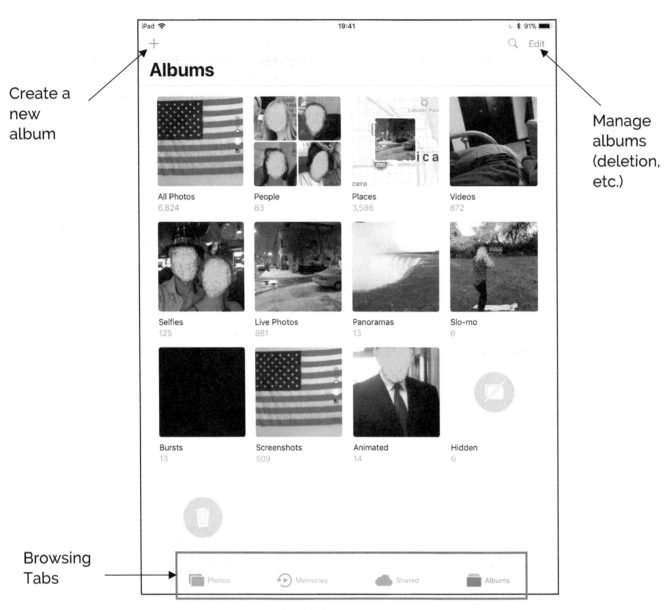

Browsing
Tabs

Figure 10.1 – Photos app -> Albums tab

iCloud Photo Library

Before we explore the Photos app more thoroughly, it is important to discuss what is called iCloud Photo Library. Photos take up a lot of space on your iPad. Videos take up even more space. Your iPad is equipped with what is called iCloud Photo Library, which basically stores your photos off of your iPad and in the cloud, which is just a hard drive that is stored on the internet. iCloud Photo Library may be a great option for you if you plan on keeping a lot of photos on your iPad. It also allows all of your photos to be easily accessed on all of your Apple devices. In order to use iCloud Photo Library your Apple ID must be set up, which we covered at the beginning of this book. If you want to enable iCloud Photo Library, follow these steps:

1. Open <u>Settings app</u>
2. Find <u>Photos</u> and tap it
3. At the top, tap on iCloud Photo Library to the right to enable it (green will show).
4. Choose whether you want to <u>Optimize iPad Storage</u> or Download and Keep Originals. (Optimize storage will save you space)
5. You will now be using iCloud Photo Library.

It may take some time for your iPad to upload all of your photos into the cloud.

Browsing Photos

Back inside the <u>Photos app</u>, let us browse around. The two best tabs at the bottom for browsing photos are <u>Photos</u> and <u>Albums</u>. If you tap on the <u>Photos</u> tab, all of your photos will be sorted by date. You can use the <u>back arrow</u> at the upper left to expand your view. (<u>Figure 10.2</u>)

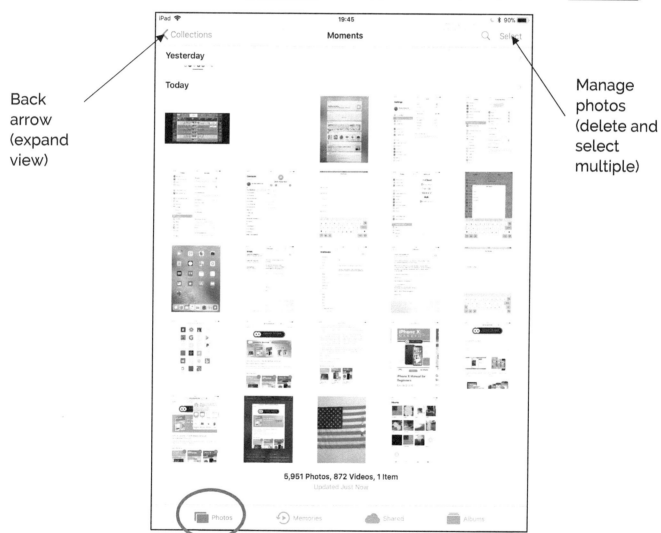

Back arrow (expand view)

Manage photos (delete and select multiple)

<u>Figure 10.2</u> – Photos app -> Photos tab

Tapping on the <u>Albums</u> tab, you can browse all of your photos as thumbnails by tapping on the <u>All Photos</u> album. To view a photo, tap on *it* to bring it full screen. Use the <u>back arrow</u> at the upper left to go back. (<u>Figure 10.1</u>)

To Create an Album

(See <u>Figure 10.1</u>)

1. Tap the <u>Albums</u> tab in Photos
2. Tap the <u>plus sign</u> at the upper left
3. Type in your new album name and tap <u>Save</u>
4. Scroll through your photos and tap each one you want to add to the album
5. Once you have selected all the photos you want, tap <u>Done</u> at the upper right

To Edit an Album

1. Open the album you want to edit in the <u>Albums</u> tab by tapping on *it*
2. Tap <u>Select</u> at the upper right
3. Tap each photo you want to manage within the album
4. Use the options at the upper left of your screen.

Sharing a Photo

(See <u>Figure 10.3</u>)

To share a photo, follow these steps:

1. Find the photo you want to share, and tap on *it* to bring it full screen. (Note: You can select multiple photos by tapping <u>Select</u> at the upper right).
2. Tap the <u>rectangle with the arrow in it</u> at the upper right.
3. Tap on the corresponding option as to how you want to share it. (<u>Message</u>, <u>email</u>, <u>Facebook</u>, etc.)

Figure 10.3 – Sharing a photo

Deleting Photos

1. Tap on the photo you want to delete (or select multiple by tapping on <u>Select</u> at the upper right)
2. Tap the <u>trash icon</u> at the upper right or left to delete

Editing a Photo
(See Figure 10.4)

Editing
options

Figure 10.4 – Edit a photo menu

You can edit a photo in multiple ways, here's how:

1. Tap on the photo you want to edit to bring it full screen.
2. Tap the Edit at the upper right of your screen.
   a. You can crop the photo by tapping on the crop icon at the bottom of your screen. Then use your fingers and drag the border to determine the crop. You can rotate the photo using the block and arrow icon. Tap the crop icon again to return. (Figure 10.5)

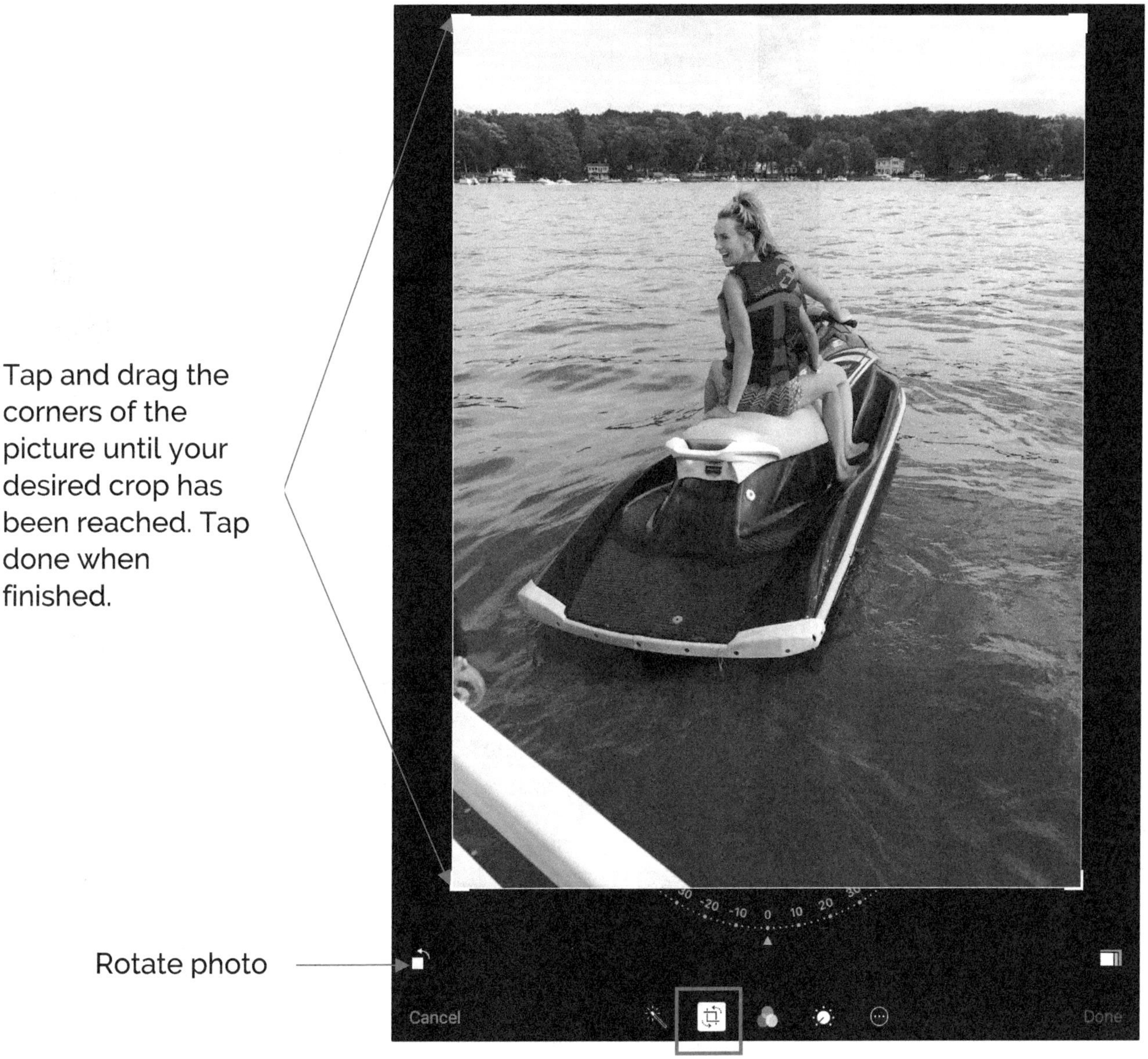
Tap and drag the corners of the picture until your desired crop has been reached. Tap done when finished.
Rotate photo
Cancel
Done

Figure 10.5 – Cropping and rotating a photo

b. You can add a filter to the photo by tapping on the color palette icon. (Figure 10.6)

Tap on a filter choice to add the filter. Tap Done when finished.

Figure 10.6 – Adding a filter to a photo

c. You can alter lighting and shading of the photo by tapping on the <u>sun icon</u>. (<u>Figure 10.7</u>)

Drag the slider
to the desired
setting

<u>Figure 10.7</u> – Adjusting lighting and color settings of a photo

d. You can markup the photo by tapping on <u>the 3 dots within the circle icon</u>. Tap <u>Markup</u>. (<u>Figure 10.8</u>)

Using markup, you can write directly on the picture by placing your finger on the photo and drawing with your finger.

Markup Options

Figure 10.8 – Marking up a photo

e. You can remove red-eye by tapping on the <u>eye icon</u>, which will appear in the editing options when your iPad detects eyes in the photo.

f. Lastly, you can use the auto enhance feature, which is the <u>magic wand icon</u>.

3. Tap <u>Done</u> on your screen when you are done editing the photo.

Photo Streams

A photo stream allows you to share a select set of photos with your friends who use Apple devices. To create a new Photo Stream:

1. Tap the <u>Shared tab</u> at the bottom inside the <u>Photos app</u>

2. Tap the <u>back arrow</u> at the upper left

3. Tap the + sign at the upper left
4. Name your shared album
5. Tap Next in the box
6. Enter in the contacts you want to share the album with one at a time. NOTE: They must be Apple Device users.
7. Tap Next when done

Chapter 11 – iPad Security

Security is a very important feature of the iPad. As I emphasized early on in this book, it is extremely important to remember your lock screen passcode. If you forget this passcode, it is extremely tedious to get into your device. Doing so actually involves resetting your device to factory conditions, which will delete all of your data. In this chapter I will show you how to change the security settings of your iPad, including your lock screen passcode. Also, please remember your lock screen passcode is different from your Apple ID password.

Settings

All of your security settings can be viewed and changed inside the Settings app. We have used the Settings app several times already in this book so you should have no problem by now opening the Settings app. As a reminder, the Settings app can be found on your home screen, and you can open it by tapping on it with your finger.

Here inside the Settings app is where you can alter all the under the hood aspects of your iPad. We will explore more in Settings later on, but for now, locate Touch ID & Passcode in Settings and tap on it. (Note: iPads that do not have Touch ID capability will just show Passcode, you also may have to look inside General inside Settings to find it).

To get into Touch ID & Passcode you will have to enter your current Passcode. You may have set one up when you first setup your iPad. Enter in your passcode and you will be brought to a new screen.

In this new screen you can alter many settings. You can choose whether to use Touch ID, change your passcode, or even turn your passcode off. Let's explore these. (Figure 11.1)

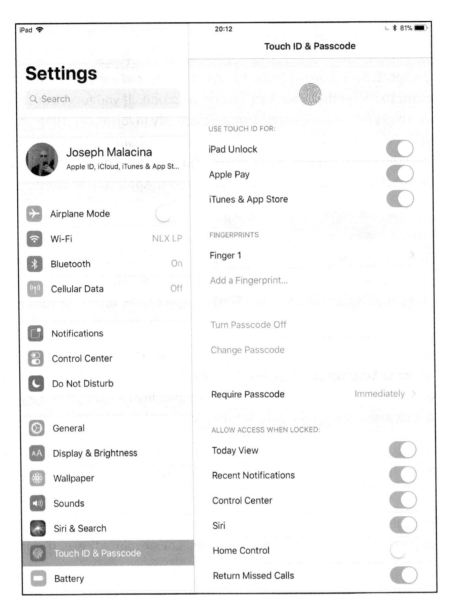

Figure 11.1 – Settings -> Touch ID & Passcode

Touch ID (if Equipped)

Touch ID allows you to use your fingerprint to unlock your iPad. It also allows you to use your fingerprint for some other security measures, such as confirming a purchase of an app or song. To turn on and use Touch ID, if you have not done so already when you first setup your iPad, follow these steps. (If you already use Touch ID, you can use these steps to add an additional finger). From within Settings -> Touch ID & Passcode (If Touch ID does not appear, then your iPad is not Touch ID compatible):

1. Tap Add a Fingerprint
2. Now follow the instructions on the screen
3. Once it's done, tap Complete/Continue at the bottom.

4. Your fingerprint will now be valid to unlock your device.

*Please note that when you use Touch ID, you will still be required to have a passcode set as well. This passcode will be required whenever your restart your iPad or if Touch ID fails.

Passcode

Your passcode is a password that allows you to unlock your iPad. Unlocking is simply the process of getting into your iPad to use it. You should use a Passcode for your iPad. Otherwise, anyone will be able to get into your iPad and have access to all of your data should they ever get possession of it.

Setting or Changing a Passcode

1. Within Touch ID & Passcode in Settings, tap Turn Passcode On or Change Passcode.
2. If you are changing your passcode, you will be required to enter your current passcode. Do so.
3. Now you can enter a new passcode. The standard is a 6-digit passcode, however many people prefer to use 4-digits or even an alphanumeric code. I personally prefer to use 4-digits. To change the type of passcode you want, tap on Passcode Options.
4. Now you can choose which type of passcode you want.
5. Tap in your new passcode.
6. Confirm the new passcode by tapping it in again.
7. Follow the instructions on the screen to finish entering your passcode.
8. When complete, your passcode will be changed or set.

To Turn Off Passcode Completely

1. Within Touch ID & Passcode in Settings, tap Turn Passcode Off.
2. Tap Turn Off
3. Enter in your current passcode.

As a final reminder, you can use your passcode to unlock your iPad from the sleep state. You can also use your fingerprint to unlock your iPad from the sleep state if your iPad has the capability. To unlock your iPad from the sleep state, first press the home button, then place your finger on the home button to use Touch ID. This will unlock your device and bring you to your home screen. If you do not have Touch ID, you will instead be prompted to enter your passcode once you press the home button.

And as a final, final reminder, please remember your passcode. I highly suggest writing it down somewhere safe in case you ever forget it. I cannot stress enough how much of a headache it can be if you forget this passcode.

Chapter 12 – Personal Settings

There are a number of personal settings you can set on your iPad, and in this chapter we will explore some of them.

Setting your Wallpapers

Your wallpaper is just your background image on your iPad. You have 2 different wallpapers: your home screen and your lock screen. To set your wallpapers follow these steps:

1. Open the <u>Settings app</u>
2. Tap <u>Wallpaper</u>
3. Tap <u>Choose New Wallpaper</u>
4. You can now choose between Dynamic, Stills, and one of your photos. Dynamic wallpapers are animated on your screen. Stills are not animated and are pre-loaded on the iPad. You can select one of your current photos by browsing through your albums at the bottom. (<u>Figure 12.1</u>)

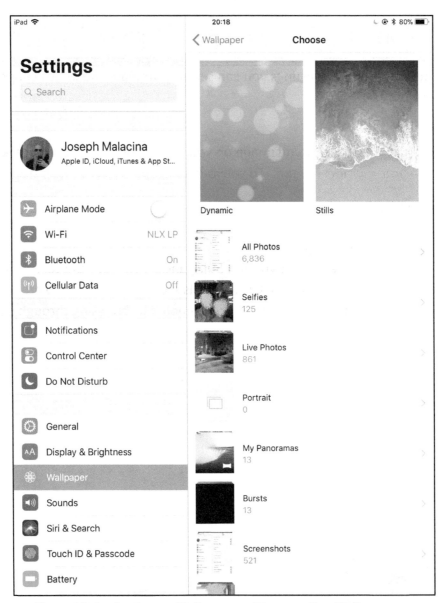

Figure 12.1 – Settings -> Wallpaper -> Choose a New Wallpaper

5. Now select a photo or image to use as your wallpaper.
6. You can adjust the Perspective Zoom if you wish. Enabling this adjusts your wallpaper to which angle you are viewing your iPad at.
7. Choose to set the wallpaper as your <u>lock screen</u>, <u>home screen</u>, or <u>both</u>.

Ringtones & Sounds

You can customize your ringtones and sounds in multiple ways. Here's how:

1. Open <u>Settings</u>
2. Tap <u>Sounds</u> (Or <u>Sounds & Haptics</u>)
3. Here you can adjust various sound settings such as ringer volume. To change a specific

tone, such as your ringtone, tap on <u>Ringtone</u>.

4. Now you can tap on each tone to preview it on your iPad. Once you have found the tone you want, use the <u>back arrow</u> at the upper left to go back.

Other Settings

You can adjust other personalized settings from within the Settings app. We will not cover them here in this text but feel free to check out <u>Display & Brightness</u> and <u>Battery</u> settings.

Chapter 13 – The Home Screen

By now, you should be pretty familiar with your home screen. As a reminder, your home screen is the main screen of your iPad that shows all of your apps. You have multiple home screens, and you can switch between them by tapping down with your finger on the screen and swiping left or right. To get back to the main home screen at any time, simply press the home button. (See Figure 4.1 in Chapter 4 as a reminder)

Organizing Apps

On your home screens, you can choose where you want each app to appear. You can also group multiple apps together in a folder.

Moving Apps Around on your Home Screen (Figure 13.1)

1. Tap and hold down on an app on your home screen until all of your apps start to shake, then release your finger. You may see a popup window while holding on an app; keep holding until the apps start to shake.
2. Now that your apps are shaking, you are allowed to move the apps around.
3. To move an app, tap down on a shaking app and do not let go. You can now drag your finger anywhere on the screen to move the app to a new location.
4. To move the app to a new home screen, drag the app all the way to the right or left edge and hold it there, until the home screen changes. Press the home button to return.

Grouping Apps into Folders (Figure 13.1)

1. Tap and hold down on an app on your home screen until all of your apps start to shake, then release your finger.
2. If you want to create a folder of apps on your home screen, tap down on an app and hold it, then drag it on top of another app that you want to group with the app you are holding.
3. You will see a small window appear behind your app. Release your finger.
4. A new folder has just been created. Your iPad will name the folder for you. To change the name of the folder, tap into the box and use your keyboard to enter in a name.
5. Tap out of the folder to return to the home screen with your apps still shaking.
6. You can organize that folder the same way you can organize apps. You can move more apps into that folder by tapping, holding, and dragging apps into the folder. Press the home button to return.

How to Delete Apps (Figure 13.1)

1. Tap and hold down on an app on your home screen until all of your apps start to shake, then release your finger.
2. To delete an app, tap the small (x) that appears at the upper left of an app.

3. Then tap <u>Delete</u>.
4. To delete a folder, just move all the apps inside a folder out of that folder.
5. Please note that some apps that came with the iPad cannot be deleted.
6. Once an app is deleted, you can always get it back in the App Store, which is covered in the next chapter.

Tap the small "x" to delete the app

Drag and drop with your finger one app to inside another app to create a folder on your home screen.

Drag and drop an app with your finger to another location on your home screen to move it there

Drag with your finger an app to the edge of your screen to move it to another home screen

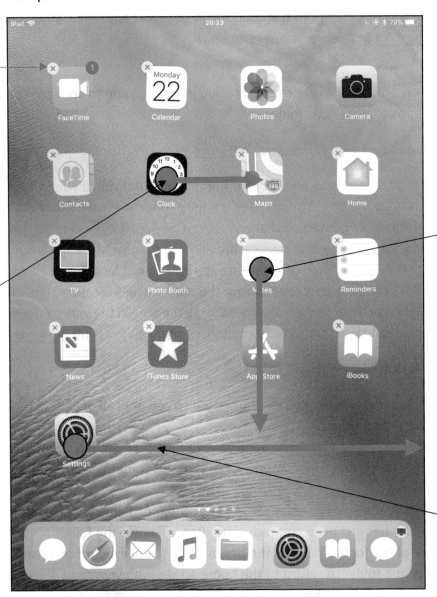

Figure 13.1 – Managing & organizing apps

Chapter 14 – Apps

All the things that your iPad can do are functions of apps. When you take a photo, you are using the Camera app. When you send a text message, you are using the Messages app. When you send an email, you are using the Mail app. Even when you are changing your settings, you are using the Settings app. The apps that come with the iPad are extraordinarily useful, but these are just a blade of grass in a prairie compared to the vast amount of apps available. To get new apps, you will need to use an app called the App Store, which is located somewhere on

your home screen.

App Store

Let's open the <u>App Store</u> by tapping on it on our home screen. In order to use the App Store, you must have an Apple ID and you must be signed in to your Apple ID on your iPad. If you still have not set up your Apple ID yet, I urge you to do so. See Chapter 5 on setting up your Apple ID.

Browsing the App Store

(See <u>Figure 14.1</u>)

Here in the App Store you can browse through the massive library of apps available for download. Again, make sure you are signed in with your Apple ID, and make sure you remember your Apple ID password as you will need it to download apps.

Here at the main page of the App Store, called the <u>Today</u> page, you can see all the new and featured apps that the App Store recommends. Let's take a look at the layout of this page, and the App Store itself, particularly the tabs at the bottom.

- **Today** – The Today tab shows you exactly what you see now, featured apps and content.
- **Games** – This tab allows you to browse gaming apps.
- **Apps** – This tab allows you to browse through all apps available.
- **Updates** – Shows you if any updates are available for apps.
- **Search** – Allows you to search for a specific app or function of an app.

Tap on an app to learn more about it

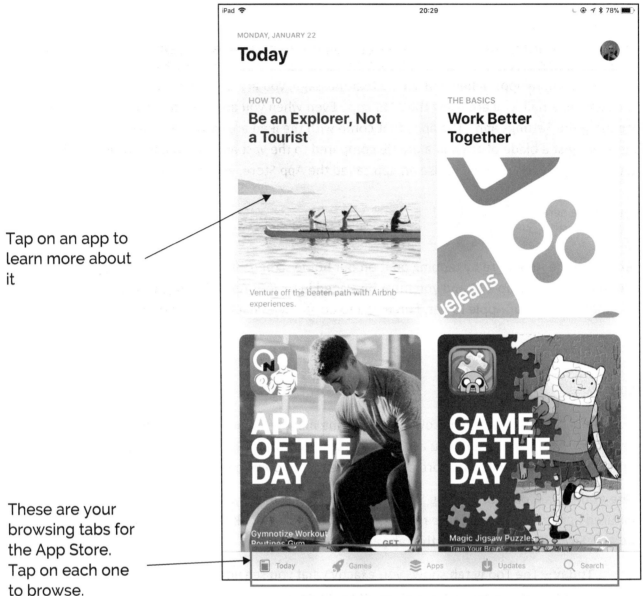

These are your browsing tabs for the App Store. Tap on each one to browse.

Figure 14.1 – The App Store -> Today tab

Browsing By Category
(See Figure 14.2)

One of the best ways to find good apps is to browse by category. To do this, first tap the Apps tab at the bottom, then scroll down until you see Top Categories. Some categories will be shown and you can tap on one to explore it. For now, tap on See All to view all app categories.

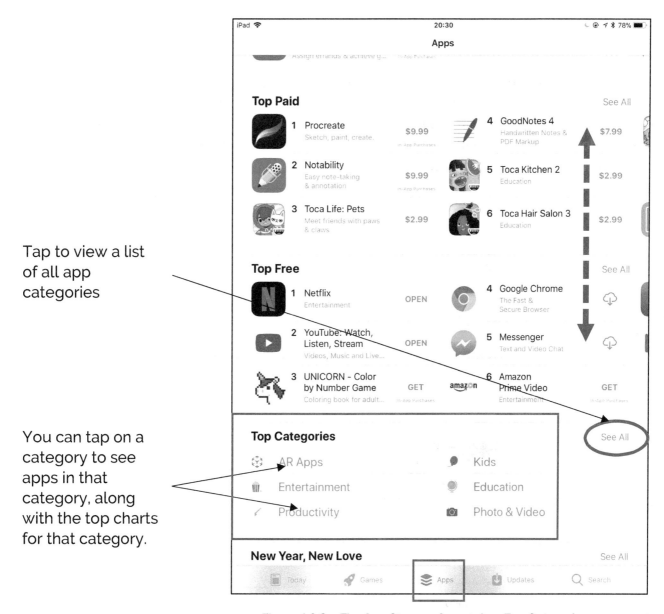

Tap to view a list of all app categories

You can tap on a category to see apps in that category, along with the top charts for that category.

Figure 14.2 – The App Store -> Apps tab -> Top Categories

When viewing all categories, you can scroll up and down to see the full list. Simply tap on a category to see apps within it. Within each category page, you will see Apps that are recommended by curators and the top charts for that category. The top charts include the most downloaded and used apps in that category and they are sorted by free apps and paid apps. Paid apps are apps you have to pay for in order to download. As always, you can use the back arrows at the upper left to go back.

Viewing an App
(See Figure 14.3)

When you see an app that interests you, you can tap on the app itself to view more information. On this screen you will see the name of the app, the price to download it if there is one, and you will see the app's user ratings, rankings, age recommendation, and more. By scrolling down and scrolling left and right you can see screenshots of the app, read a description of the app, and read user ratings and reviews.

User ratings (average star rating), app rank, & age recommendation

Tap this box to download, install, or open the app, depending upon if you already have the app installed

You can scroll up and down & left & right to see more information about the app including screenshots, description, and reviews

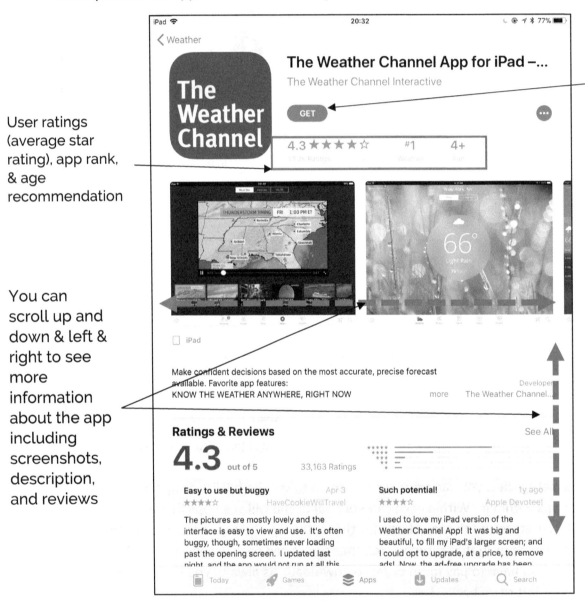

Figure 14.3 – Viewing an app in the App Store (Example: The Weather Channel App)

Top Charts

Using Top Charts is one of my favorite ways to find popular apps. We have already demonstrated how to view Top Charts for categories, but you can also see Top Charts for all apps regardless of category. To access these Top Charts, first make sure you are on the Apps tab. Then, scroll down until you come upon the headlines **Top Paid** and **Top Free** (<u>Figure 14.2</u>). These two sections show you the most downloaded and used paid and free apps. You can tap on <u>See All</u> to see the full list (<u>Figure 14.4</u>), and you can tap on any app to learn more information.

Tap on an app for more information

Tap $PRICE to purchase and download a paid app

Tap GET to download a free app.

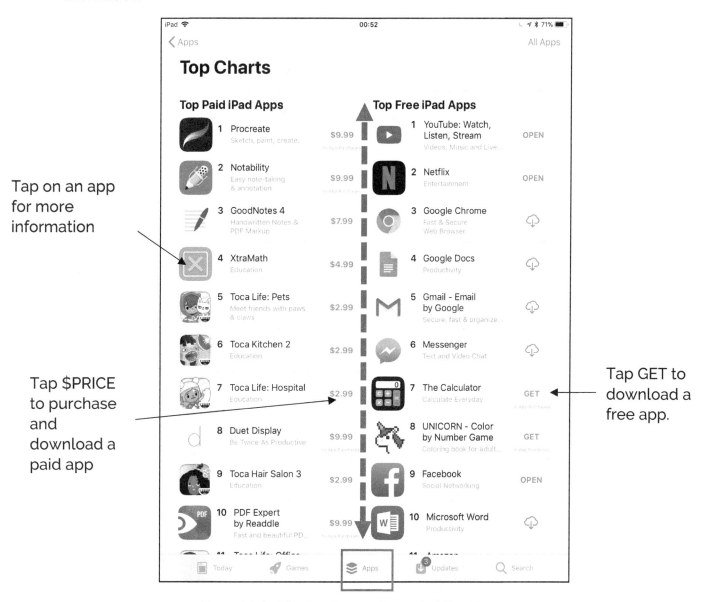

<u>Figure 14.4</u> – The App Store -> Apps tab -> Top Charts

Searching for Apps

If you know the specific app you want to download, you can search for it using the <u>Search</u> tab at the bottom. Tapping on this tab will allow you to search for any app by tapping into the <u>search bar</u> at the top of the screen. Simply type in the name of the app, or a subject such as "photo" and tap <u>search</u> on your keyboard. Usually, when you type the search term into the search field the App Store will make suggestions as to what you are looking for. You can tap on that suggestion to quickly enter it in and search. Now you will see a list returning your search results, and you can view these apps by tapping on them.

Updates

The <u>Updates</u> tab shows you what updates, if any, are available for apps you already own. You can tap <u>Update</u> next to the app to install the update. Alternatively, you can tap <u>Update All</u> at the upper right to install all available updates. There is also a way to set your iPad to update all your apps automatically. This is covered in the Tips & Tricks chapter.

Downloading Apps

Now that we have seen how we can browse apps, let us see how we actually download one to our iPad. The process is actually pretty simple, but before we begin let me give a note on your Apple ID.

As emphasized early in this chapter, in order to use the App Store you must be signed in with your Apple ID. You are also going to need your Apple ID password to download apps. On occasion, your iPad will ask for this password just for verification purposes. Lastly, you will need credit card information linked to your Apple ID in order to download certain apps. Adding your credit card information is fairly simple. Your iPad will ask you when it needs this information, and will bring up a screen with directions. Adding your credit card data does not necessarily mean you will be charged. Apple just needs it IN CASE you decide to purchase something.

Here is how you download an app:

1. Open the <u>App Store</u>
2. Find the app you want to download
3. Next to the app will be a box. Inside this box will be the words <u>GET</u> or a price such as $0.99. (See <u>Figure 14.4</u>)
4. Tap this box.
5. Next tap the same box again. This time the words <u>INSTALL</u> may be in the box.
6. You may be asked for your Apple ID password at this point. Enter it in. Alternatively, you may be asked to place your finger on your home button for Touch ID.
7. You may be asked to update your billing information at this point, such as credit card and billing address. Enter it in if prompted.
8. Your download will now commence. When it is finished, your newly downloaded app will appear on your home screen somewhere.

Using Downloaded Apps

When you download an app, it will appear on your home screen. To open it, just tap it. Using an app from the App Store is the same as any using any app that comes with the iPad. You can usually scroll through screens on the app, tap icons, and more. Each app is unique to itself, so you will have to play around with the app to get the hang of it.

There are certain aspects of apps that are important to consider. The first is *in-app purchases*. Many apps allow you to buy things from directly inside the app. These purchases could be access to content or even physical goods. To make these purchases you will most likely using your Apple ID. Another aspect is advertisements. Some apps have advertisements running through them. For most apps these are non-conspicuous and do not interfere with using the app, but for others the advertisements can be quite bothersome. If an app is flooded with advertisements and popups that make it difficult to even use, I would consider deleting the app and using something else. Apps like these are just trying to spam you with advertisements rather than provide a good service.

It is also worth noting that some apps offer a paid and free version. Usually the paid version comes with extra features or no advertisements.

Resources for learning how to use Apps

You can check out the appendix of this text for a list of popular and useful apps. You'd be amazed at what types of apps are out there and what they can do.

Some apps are a little tricky to learn, and there are a plethora of resources out there to help you learn. For learning how to use certain apps, we recommend checking out www.infinityguides.com. That website is dedicated to teaching people how to use many popular apps from a beginner's perspective.

Chapter 15 – Notifications

So far we have covered all the basics of using the iPad. You now know how to navigate the iPad, browse the Internet, use your email, personalize your device, download apps, and more. Now we are going to get into specific features that are absolutely essential to understand, and we start with notifications.

Overview

Notifications are an integral part of your iPad. Each time certain information is delivered to you from an app, you will receive what is called a notification. An example of this would be a breaking news alert from a news app or a notification that you received a new email.

Notifications appear on your iPad in several forms. On your lock screen, you will see notifications you have missed lined up (Figure 15.1).

These are notifications

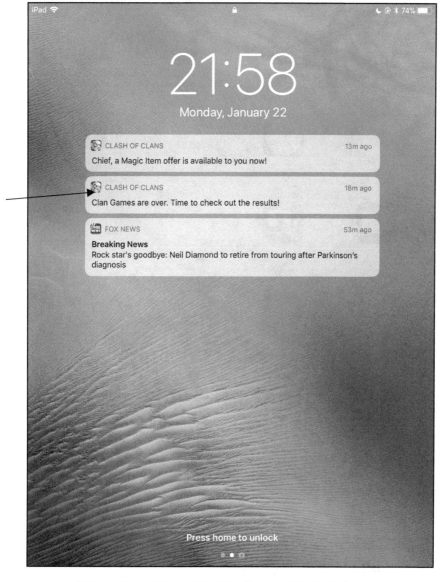

Figure 15.1 – Missed Notifications on the Lock Screen

While using your iPad, you will see notifications in real-time pop up at the top of your screen. (Figure 15.2). When these notifications appear, you can tap on it to be brought directly to the notification. For instance, if you receive a text message while you are doing something on your iPad, a box will appear at the top of your screen showing the text message and who it is from. This box will appear briefly. While the box is there, you can tap on it to be brought directly into the thread of the text message so you can respond. Alternatively, you can tap that box and drag your finger down to reply immediately. I cover texting on your iPad in Chapter 20.

Another type of notification is the badge icon, which is simply the red bubble you can see above an app on the home screen that has a number in it. This usually means you have a certain number of notifications available for that particular app. For instance, your Mail app will show a number in a red bubble above it on your home screen. This number indicates how many unread email messages you currently have.

Real-time notification. Tap the horizontal line and drag down to see the full notification

Badge icon, usually indicates number of notifications for that app, in this case the number of unread text messages

Figure 15.2 – Real-time Notifications while iPad unlocked

Notifications can also appear in real-time as an alert. An alert is larger box that appears in the middle of your screen and they usually require you to dismiss the notification before it goes away. An example of this is a severe weather alert or an Amber alert.

The Notification Center

At any time you can view all of your current notifications by accessing the Notification Center. To do this, tap down at the very top of your screen at any time and swipe your finger down. This is your Notification Center (Figure 15.3). Here you can see all the notifications that you have not acknowledged. To go to a particular notification, simply tap on it. To clear out a section of notifications, tap the small x next to the date, and then tap clear. Just press the home button to return to your home screen.

To open the Notification Panel, tap down at the very top of your screen and swipe down

Tap to clear the particular set of notifications

Figure 15.3 – The Notification Center

Examples of Notifications

As stated earlier, a notification can be from any app. For instance, a breaking news story from Fox News can be a notification if you have the Fox News App. A new Facebook like on your post can be a notification if you have the Facebook app installed. Basically, any app can deliver notifications.

Managing Notifications

Sometimes, you may not want to get notifications from a certain app. These can be easily managed in the Settings app.

1. Open the Settings App
2. Tap Notifications
3. Tap the app you want to manage notifications for
4. Now you can choose whether you want to allow notifications from this app. You can also choose the type of notification as well as some other options. (Figure 15.4)

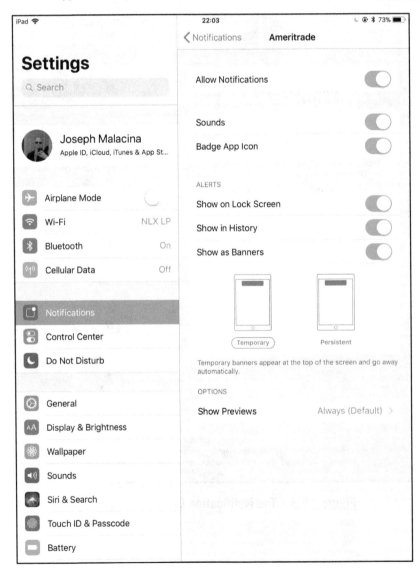

Figure 15.4 – Settings -> Notifications -> App Name (Ameritrade)

Chapter 16 – The Control Center

The Control Center, sometimes referred to as the "control panel," is a nifty tool on the iPad that allows you to quickly change some settings. Let's explore it.

To Access the Control Center

1. On any screen on your iPad, tap down at the very bottom of your screen and swipe all the way up. You will need to swipe up to at least the halfway point of your screen. (If you only swipe up a little, you will bring up your home screen dock.) You will now see the Control Center and Recently Used Apps (Figure 16.1). (Alternatively, you can double press the home button and this will accomplish the same thing).

Airplane mode

Enable/disable cellular data

Turn Bluetooth on/off

Turn Wi-Fi on/off

Music Controls

Adjust brightness

Adjust volume

AirPlay

Lock/Unlock Orientation

Turn Mute on/off

Open Timer (Clock App)

Turn Do Not Disturb on/off

Open Camera App

To open the Control Center, tap down at the very bottom of your screen and swipe up

Figure 16.1 – The Control Center

Control Center Functions

At the top right side of your Control Center you will see some buttons. Tapping these buttons turns the selected setting on or off.

- **Airplane Mode** – Tapping on the airplane button turns Airplane Mode on or off. When a button is highlighted that means it is currently enabled. Airplane Mode turns off all communication abilities on the iPad, including cellular, Wi-Fi, and Bluetooth. Once you turn Airplane Mode on, you can still enable Wi-Fi mode by using the Control Center.
- **Wi-Fi** – Underneath airplane mode, tapping the Wi-Fi button will turn Wi-Fi on or off.
- **Cellular Data** – The antenna icon enables or disables cellular data.
- **Bluetooth** – The Bluetooth icon turns Bluetooth on or off. To connect to a Bluetooth device, you need to go to Settings -> Bluetooth.

Underneath these four button will be the Music controls, which let you quickly control your music. Underneath the Music controls will be the brightness and volume sliders. You can adjust these by tapping in the slider and sliding your finger up or down.

Underneath the buttons and brightness slider will be some tabular buttons:

- **Screen Mirroring** – This button lets you broadcast your iPad's screen to an AirPlay enabled device, such as an Apple TV.
- **Rotation Lock** – The rotation lock button will lock the current orientation of your iPad. When this is enabled while viewing your iPad in portrait orientation, turning your iPad to landscape orientation will not re-orientate the screen. If your iPad screen is not orientating to landscape mode this option may be on.
- **Mute** – The bell icon allows you to turn Mute mode on or off. When Mute is on, your iPad will make no sounds for notifications. When Mute is off, your iPad will make sounds for notifications.
- **Do Not Disturb** – The moon button enables or disables Do Not Disturb mode. When you enable Do Not Disturb mode, you will not receive notifications or hear any sounds for phone calls and messages while your screen is off. These along with all other notifications besides your alarm clock will be silenced.
- **Timer/Alarm Clock** – The clock icon opens your Clock app, and brings you directly to the Timer. In the Timer, you can set a timer that will notify you when the duration elapses with an alert and sound. You can also set your alarm clock from here by tapping on the Alarm tab at the bottom. To create a new alarm, tap the plus icon at the upper right. Alternatively, you can enable a previously created alarm by tapping the small oval next to a time. You can now set the time for your alarm and specify which type of alert you want. Tap Save at the upper right to finish setting the alarm.
- **Camera** – The last icon at the bottom opens the Camera app.

- **Night Shift –** You can access Night Shift by tapping and holding on the <u>brightness</u> slider until it focuses full-screen. When this happens a small <u>Night</u> Shift button will appear under the slider (<u>Figure 16.2</u>). This button allows you to turn Night Shift on and off. When you turn Night Shift on, your iPad will change its color display. The purpose of Night Shift is for when you are using your iPad late at night and do not want your iPad's screen straining your eyes and potentially keeping you awake. Night Shift will automatically turn off the next morning.

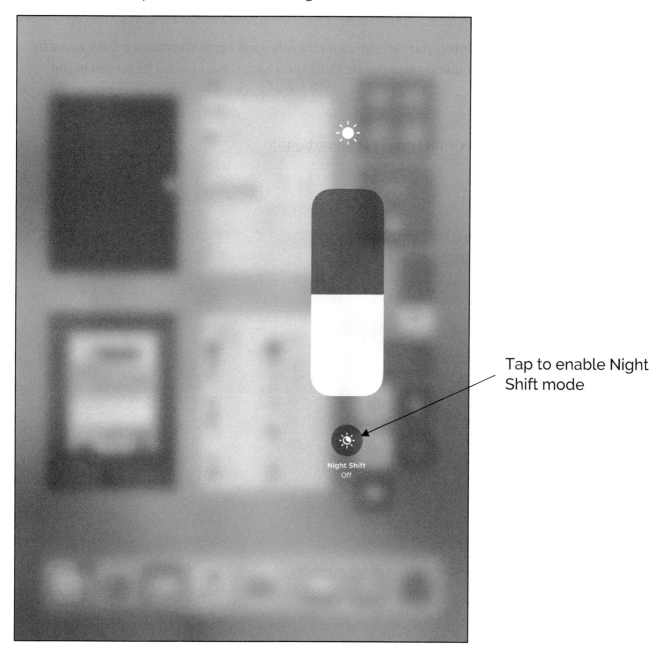

Tap to enable Night Shift mode

Figure 16.2 – Accessing Night Shift in the Control Center

Music Controls

The Music box in the Control Center allows you to quickly control your music. While a track is playing, you can bring up the Control Center and then use the skip back, pause/play, and skip forward icons to quickly navigate through your music without having to open up the Music app. You can also bring up more options such as broadcasting to a Bluetooth connected speaker by tapping on the <u>sound wave icon</u> at the upper right of the Music box. Simply tap out of the enlarged Music box to return to the main Control Center.

Recently Used Apps

Also from the Control Center, you can see your recently used apps. You can see even more by swiping left and right. To quickly access a recently used app, simply tap on its screen in the Control Center.

Exiting the Control Center

To exit the Control Center either press the <u>home button</u>.

Chapter 17 – Siri

Siri, as you may have heard, is Apple's intelligent voice assistant. You can talk to Siri, and Siri will listen to what you say and try to do whatever you command. The possibilities of what you can do with Siri are endless.

Turning Siri On
(See Figure 17.1)

By default, Siri is enabled on your iPad. To be sure, let's make sure Siri is enabled. To do this, follow these steps:

1. Open Settings.
2. Find and tap Siri & Search on the left.
3. Make sure Press Home for Siri is enabled at the top. (The slider tab next to it should be green. If not, tap it).

While we are in the Siri settings screen, let's take a quick look at some of the settings you can adjust.

- **Allow Siri When Locked –** Allows you to access Siri when your iPad is locked. I usually leave this enabled so I can access Siri quickly without unlocking my iPad.
- **Listen for "Hey Siri" –** With this enabled, you can access Siri by simply saying "Hey Siri" at any time. Personally, I do not use this option as there is a much more reliable way to access Siri.
- **Other Settings –** The settings at the bottom allow you to change some other options including which language you want Siri to use and which voice you want Siri to have.

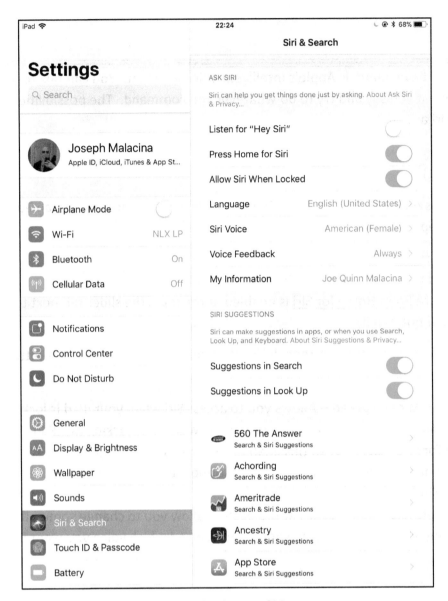

Figure 17.1 – Settings -> Siri

Accessing Siri

Now that Siri is enabled on your iPad, let's go back to the home screen. From the home screen, or any screen for that matter, accessing Siri is very easy. The easiest way to access Siri is to press and hold the <u>home button</u> until you see a sound wave appear at the bottom of your screen (<u>Figure 17.2</u>).

You can make Siri start listening again by pressing on the microphone icon, or pressing and holding the home button again

Figure 17.2 – Using Siri

Using Siri

Now let's try some examples to see exactly how Siri works. First, let me note that when you access Siri by holding the home button, you have two different approaches. One is, when the sound wave appears, you can let go of the home button and start speaking. Two is, when the sound wave appears, you can continue holding down on the home button, start speaking, and release the home button when you are done speaking. Either way works. For simplicity, I am going to use the former method in our examples.

So let's press and hold the home button until the sound wave appears then release the home button. Now we can say, "What is the weather?" Siri will listen, then respond with some information. She can either speak the information back to you or bring up a screen with the information you requested. To start a new Siri session or attempt to speak again, tap on the microphone icon that appears at the bottom of your screen or just hold the home button again.

Let's try another example, press and hold the <u>home button</u> and release, and then say "Show me my photos." Siri will listen and then automatically open the Photos app for you.

There are countless things you can say to Siri that she will perform. Try out anything you would like. A few more examples are listed below, and you can see a much larger list of examples in the Appendix of this text.

- Lookup Contact Information – "Lookup *Contact Name*"
- Lookup Sports Scores – "Who won the White Sox game today?"
- Set an Alarm Clock – "Set an alarm for 6:30 AM"
- Do Math – "What is two plus two?"
- Visit Websites – 'Bring me to Infinity Guides dot com"
- Listen to latest news (Will play NPR segment) – "Listen to the news"

Chapter 18 – Native Apps

Let's explore some native apps that come preloaded on the iPad. Many of these are very useful and you may find yourself using them often.

Music

(See Figure 18.1)

The Music app is where you can play and listen to your music. There are two ways to listen to music. The first is to load songs that you own onto your iPad. To do this, you will need to use iTunes, which is a program for computers. To see how to do this, go to www.infinityguides.com and search for "iTunes".

The second way to listen to music is to use the Apple Music service. This service lets you listen to nearly any song you want at any time, right from your iPad and/or any computer or device. The Apple Music service is huge, and is a subscription service from Apple. To learn how to use Apple Music and more, go to www.infinityguides.com and search for "Apple Music".

Once you have Music on your iPad using any of the two above methods, using the Music app is pretty simple. You can browse through the app using the tabs at the bottom, and simply tap on a song to play it. You can also quickly skip through tracks using the Control Center, which is covered in Chapter 16.

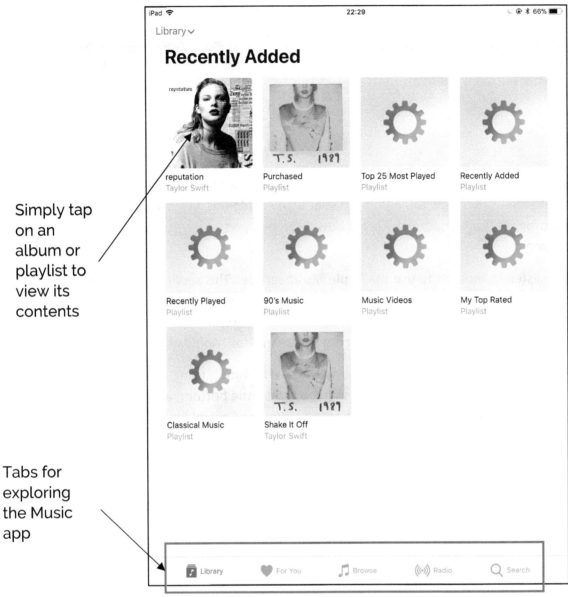

Simply tap on an album or playlist to view its contents

Tabs for exploring the Music app

Figure 18.1 – Music app

Maps

The Maps app is a great app for finding places. Inside the Maps app you can tap into the <u>search bar</u> and type in the name of a place or category and see what's around. In other words, you can look up restaurants nearby by searching for "restaurants", or you can look up movie theaters by searching for "movie theaters". You can even get turn-by-turn directions using the Maps app. To do this, tap on a selection that you searched for and then tap on <u>Directions</u> to get driving directions to the location. Your iPad will give you turn-by-turn directions as you are driving. (<u>Figure 18.2</u>)

*Please use extreme care and discretion if you choose to use your iPad for turn-by-turn directions while driving. Always be safe. Also, keep in mind that in order to use turn-by-turn navigation outside of your Wi-Fi range you will need a cellular data plan for your iPad. It is recommended that you use an iPhone with a window or similar mount for GPS directions instead.

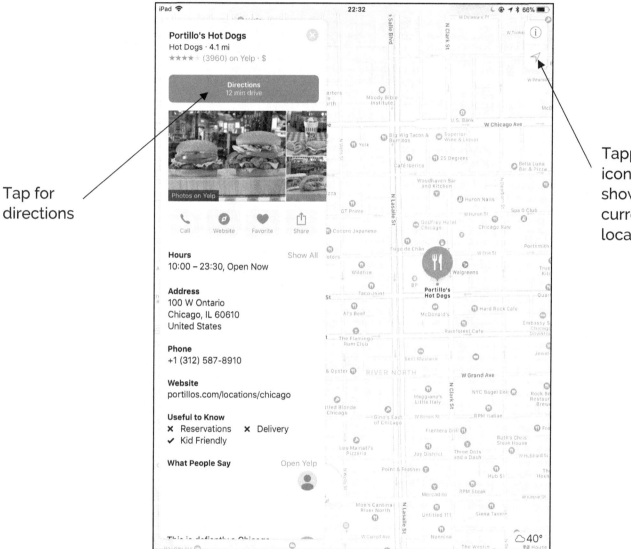

Tap for directions

Tapping this icon will show your current location

Figure 18.2 – Maps app

Note: There are other apps that offer GPS and turn-by-turn directions. Another popular app is Google Maps, available in the App Store for free.

FaceTime

(See Figure 18.3)

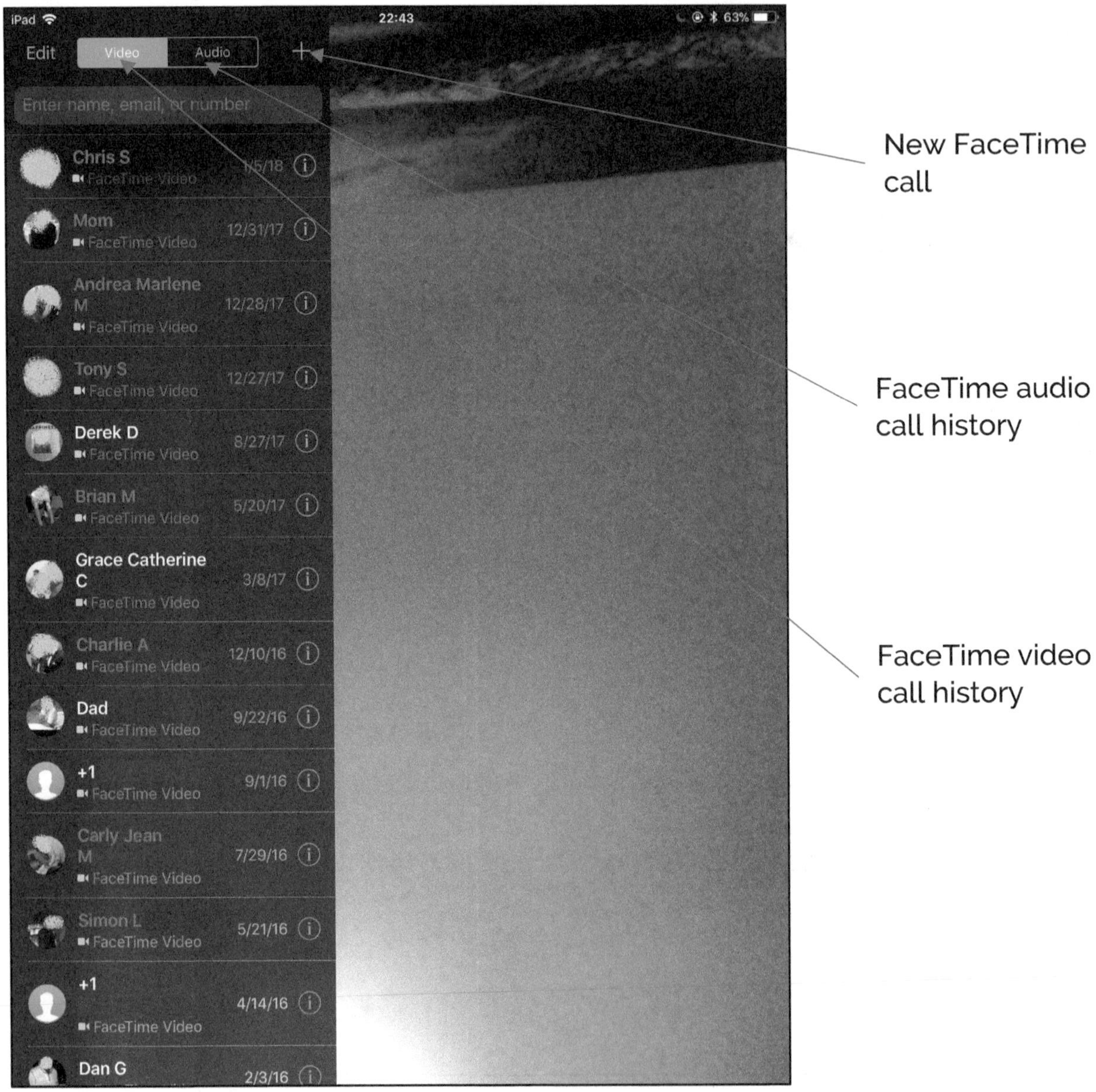

New FaceTime call

FaceTime audio call history

FaceTime video call history

Figure 18.3 – FaceTime app

FaceTime, which I have mentioned briefly earlier in this book, is an app that allows you to partake in video calls with other Apple users. This app is very similar to Skype, which is also available for free in the App Store. When you open FaceTime, you can place a video call by tapping on the plus sign at the top. Now you can look through your contacts to find someone to video chat with. When you have found that person, tap on their *name*. Now FaceTime will determine if they are able to make a FaceTime call. You will know if you are able to FaceTime that person because some icons may appear next to their name. If the video recorder icon appears, that means that you can make a video FaceTime call with them. If the phone icon appears next to their name that means you can make a FaceTime Audio call with them, which has no video. If both icons are greyed out, and you are not allowed to tap on them, that means you cannot make a FaceTime call of any type with them. In other words, they do not have FaceTime on their device. (<u>Figure 18.4</u>)

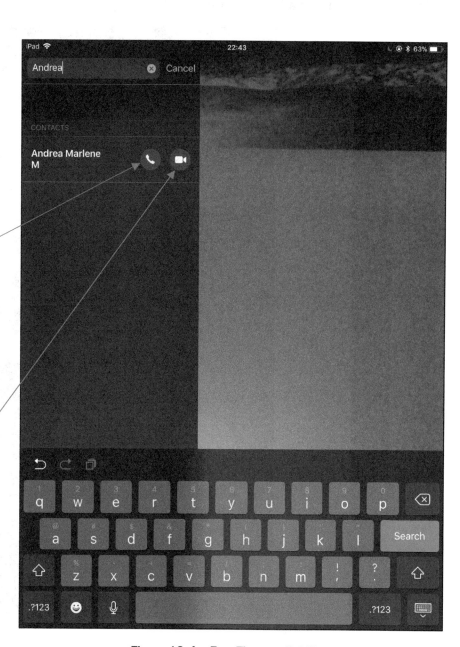

A phone icon means this person can accept FaceTime audio calls. Tap the phone icon to initiate the call.

A video recorder icon means this person can accept FaceTime video calls. Tap the video icon to initiate the call.

<u>Figure 18.4</u> – FaceTime availability

Clock

We have already covered most of the Clock app in Chapter 16. There are a few things we missed, so we will cover them now. In this app you can see the time around the world, as well as set an alarm, and start a timer. You can also use the stopwatch function by tapping on the Stopwatch tab at the bottom. Lastly, the Bedtime tab allows you to set up a custom bedtime and wake-up alarm that adjusts to how you sleep. (Figure 18.5)

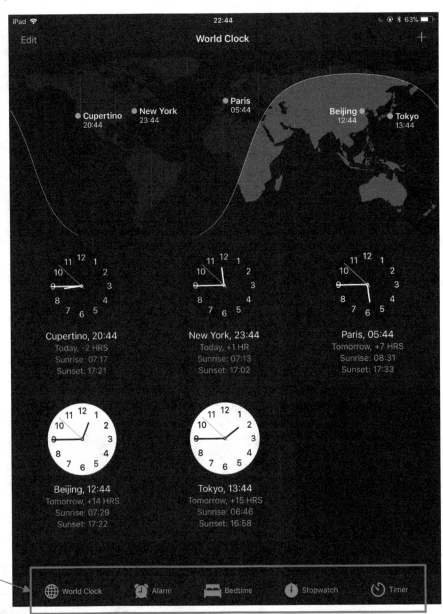

Clock tabs for different functions (alarm clock, bedtime function, stopwatch, timer, and world clock)

Figure 18.5 – Clock app

Notes

The Notes app is an incredibly useful app for creating notes. It is very easy to use; just open the Notes app, tap the <u>pencil and square icon</u> at the upper right to start a new note, and begin typing away. Notes save automatically, and you can access all of your saved notes in the main notes screen (use the <u>back arrows</u> at the upper left).   You can delete notes by tapping on <u>Edit</u> at the top.

Furthermore, while creating a note you can sketch and use additional tools by using the icons at the bottom of your screen. You can also password protect certain notes by opening the note, and tapping the <u>square and rectangle</u> at the upper right, and then tapping <u>lock note</u>. Notes save automatically. (<u>Figure 18.6</u>)

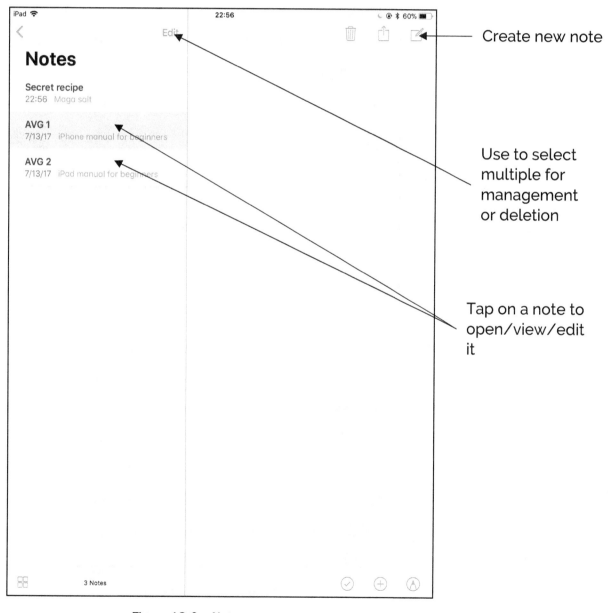

<u>Figure 18.6</u> – Notes app

iBooks

<u>iBooks</u> is the app that lets you read books on your iPad. All e-books that you download through the iBooks Store, or place in your device through iTunes will be available in iBooks.

You can also browse through books and download them using the tabs at the bottom of iBooks, these include: My Books, Featured, Top Charts, Top Authors, and Purchased (<u>Figure 18.7</u>). Finding books in the iBooks Store is very similar to using the App Store. You will need to be signed in with your Apple ID to purchase and download books. Once you purchase a book using your Apple ID, it will automatically be sent to all of your Apple devices that are signed in with your Apple ID. So if you buy a book on your iPhone, it will also appear on your iPad for easy reading. You may even be reading this book on your iPad!

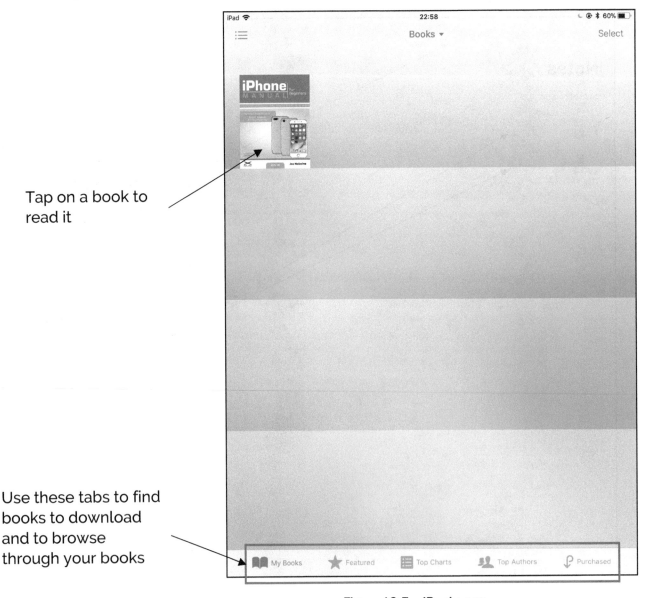

Tap on a book to read it

Use these tabs to find books to download and to browse through your books

Figure 18.7 – iBooks app

All books you download will appear in your <u>My Books</u> tab and you can read one by simply tapping on it. When reading the book, swipe your finger left or right on the page to flip through pages. (<u>Figure 18.8</u>)

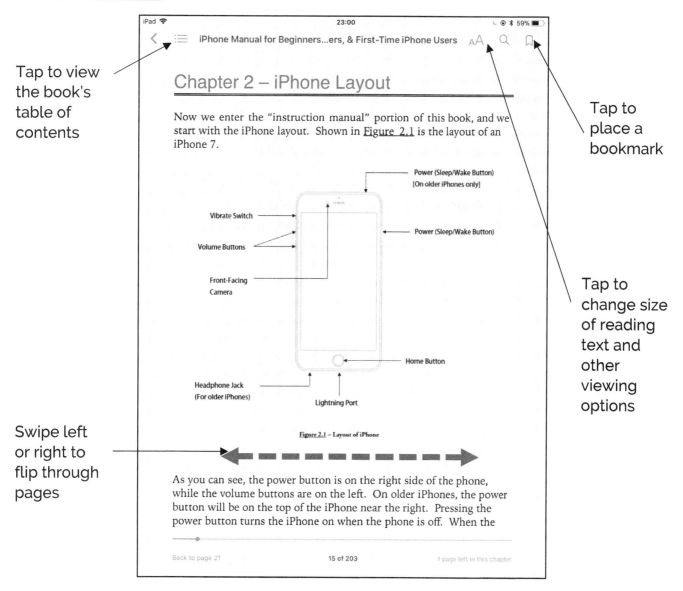

Tap to view the book's table of contents

Tap to place a bookmark

Tap to change size of reading text and other viewing options

Swipe left or right to flip through pages

Figure 18.8 – Reading a book in the iBooks app

Calendar (See <u>Figure 18.9</u>)

The <u>Calendar app</u> is a pretty simple app that allows you to create events and reminders. To view all the events you have on a day, simply tap on that day.

To Create a New Event

1. Tap the <u>plus icon</u> at the upper right.
2. Enter in all the corresponding information, if required. Include the date and time (if applicable).
3. Choose whether you want to set an alarm to remind you about this event.
4. When you are done, tap <u>Add</u> at the upper right.
5. Your event has now been added to your calendar.

You can tell if you have scheduled events on a day if text appears inside the date. You can use the tabs at the top to change your range view of your calendar.

You can also see a quick view of what you have coming up on your calendar from the home screen. Simply go to your home screen, and swipe your finger to the right until you get to the Today screen. If you have any events in your calendar coming up soon, they will appear here.

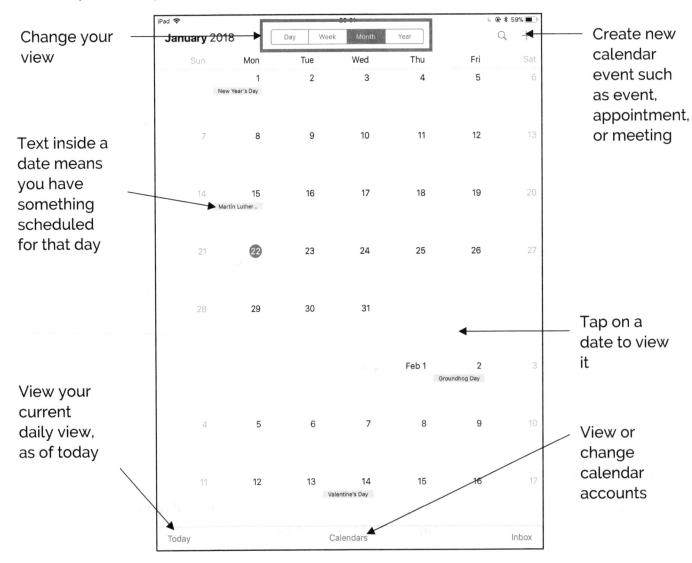

Change your view

Create new calendar event such as event, appointment, or meeting

Text inside a date means you have something scheduled for that day

Tap on a date to view it

View your current daily view, as of today

View or change calendar accounts

Figure 18.9 – Calendar app

Podcasts

Podcasts is an app that lets you listen and play podcasts, which are recorded audio and video segments. In the Podcasts app, you can browse and search for podcasts using the tabs at the bottom the same way you would in the App Store or iBooks Store (Figure 18.10). You can subscribe to podcasts, which will place that podcast in your Library tab for easy listening later on. I suggest using the Browsing tab, and then tapping on Top Charts to see which podcasts are popular. I highly recommend giving the Podcasts app a try.

Tap on a podcast to learn more about it, and to be brought to a page where you can subscribe

Use these tabs to view your subscribed podcasts and find new podcasts to listen to

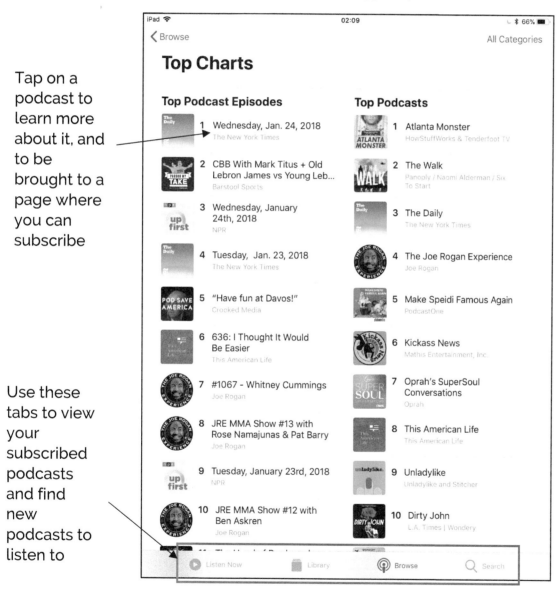

Figure 18.10 – Podcasts app -> Browse Tab -> Top Charts

Chapter 19 – iPad Phone Calls

Fun fact! The iPad can be used to make actual phone calls just like a cell phone. However, this ability is not available to everyone. In order to be able to use your iPad to make phone calls, you must have an iPhone with iOS 8 software or newer (iOS 11 preferred). **So if you currently do not own and use an iPhone, you can skip this chapter entirely.**

In this chapter I will be showing you how to set up phone calls on your iPad. This is different than using FaceTime which was covered in the previous chapter. Remember, the FaceTime app is only for making video or audio calls with other Apple device users. When you set up iPad phone calls, you can use your iPad just like you use your cellular phone. You will need your iPhone handy for this chapter, so pull it out and let's get started.

Setting up iPad Phone Calls (Using your iPhone & iPad)

For this section you are going to need your iPhone and your iPad handy. Most of the setup process is done on your iPhone.

How to enable iPad Phone Calls on your iPad (Figure 19.1)

1. First things first, update your iPhone to the latest version of iOS. You can follow the steps in Chapter 3.5 of this book for your iPhone. If your iPhone cannot upgrade its iOS to iOS 8 or newer, then you will be unable to set up and use iPad phone calls.
2. Once your iPhone's iOS is updated, you need to make sure both your iPhone and iPad are signed in to the same Apple ID. See Chapter 5 on signing in with your Apple ID on your iPhone and iPad. iPad phone calls will not work unless both devices are signed in to the same Apple ID.
3. You also must make sure that both devices are connected to the same Wi-Fi network. For more on connecting to Wi-Fi, see Chapter 4.
4. On your **iPhone**, open the Settings app on your home screen.
5. Scroll to find the Phone option and tap it.
6. Tap on Calls on Other Devices
7. Tap on the oval tab next to Allow Calls on Other Devices until the green enabled icon is shown (Figure 19.1).
8. On this screen, you can enable your Apple devices to make and receive calls by tapping on their enable option, specifically next to the name of your iPad (Figure 19.1).
9. Once enabled, a message will appear on your iPad stating that calls are now allowed.
10. Your iPad is now ready to make and receive phone calls.

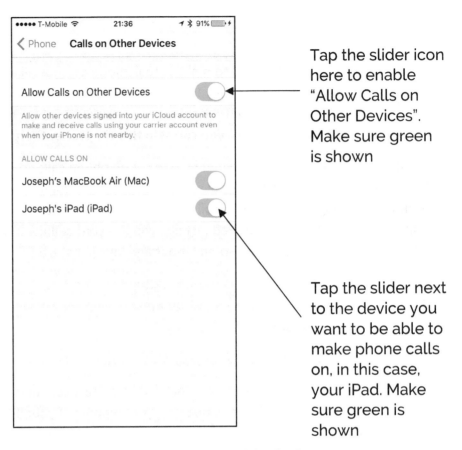

Tap the slider icon here to enable "Allow Calls on Other Devices". Make sure green is shown

Tap the slider next to the device you want to be able to make phone calls on, in this case, your iPad. Make sure green is shown

Figure 19.1 – [On iPhone] Settings -> Phone -> Calls on Other Devices

How to call someone on your iPad (Figure 19.2)

In order to call someone on your iPad, a couple conditions must be met. These are:

a. Your iPhone and iPad are connected to the same Wi-Fi network *or* your cellular service allows for Wi-Fi calling (in which case your iPhone does not necessarily need to be on same Wi-Fi network)
b. You must be signed in with the same Apple ID on both your iPad and iPhone.
c. The person you want to call is a contact in your contact list.

If all these conditions are met, then you can make a phone call on your iPad. Here's how (**on iPad**):

1. Open the Contacts app
2. Find the contact you want to call and tap on their name to bring up their information.
3. Tap on the call icon or their phone number
4. Your iPad will connect to your iPhone and then proceed to connect the call.

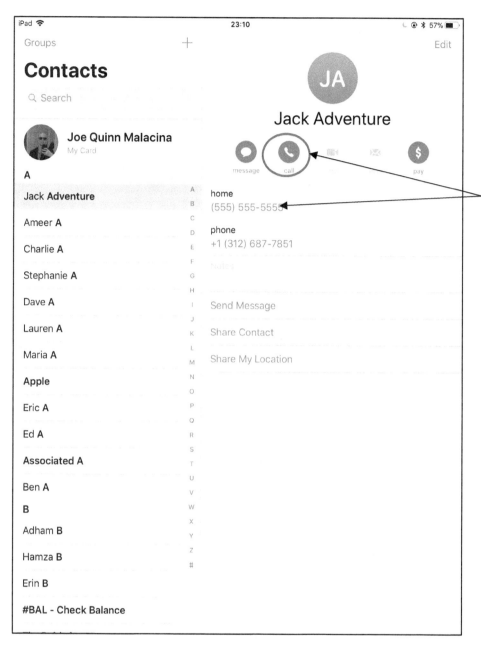

Tap on the blue call icon or on the contact's phone number to initiate the call

Figure 19.2 – [On iPad] Contacts app – Making a phone call

Receiving a Call

If you have enabled iPad phone calls on your iPad, when you receive a call on your iPhone you will be able to answer it from your iPad. When you are a receiving a call, your iPad will ring and the following screen will appear (See Figure 19.3). If your iPad is currently unlocked (i.e. your iPad screen is on), all you need to do to answer the call is tap on the green phone icon. When you do, you will now be connected to the call.

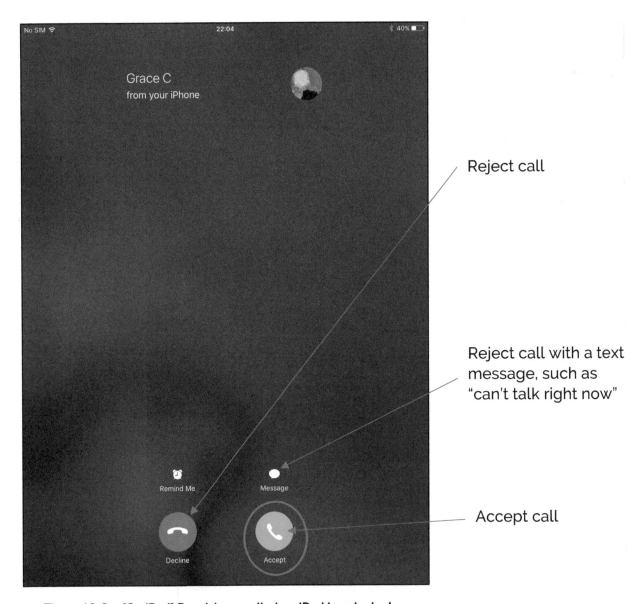

Reject call

Reject call with a text message, such as "can't talk right now"

Accept call

Figure 19.3 – [On iPad] Receiving a call when iPad is unlocked

When you receive a call when your iPad is currently locked, you will notice a slider appearing at the bottom of your screen. (See Figure 19.4). To answer the call, touch down on the green phone icon, and slide it to the right with your finger, releasing at the end. This will connect you to the call.

Tap on green phone icon and slide finger to right and release to answer

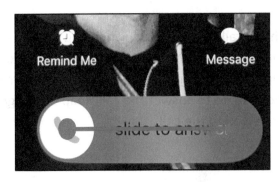

Figure 19.4 – Receiving a call, iPad locked

You can always reject a call by pressing the <u>power button</u> on your iPad when a call is incoming.

Call Functions

During a phone call, you have several functions available to you. Looking at your screen during a call (<u>Figure 19.5</u>), there are several buttons on your screen. They do the following if you tap on them:

- **Mute** – Mutes your side of the call. In other words, the person you are talking to on the phone will not be able to hear anything from your side until you touch <u>Mute</u> again.
- **Keypad** – Brings up a touch-tone keypad, in case you need to work through a touch-tone system.
- **Speaker** – Tapping <u>speaker</u> turns on speakerphone. Tap it again to turn it off.
- **Add Call** – Tapping on <u>add call</u> lets you add another person to the call, such as 3-way calling. (If part of your service).
- **FaceTime** – FaceTime lets you connect to the person you are talking to over a FaceTime call, which is a video call. This only works when you are talking to someone who has a FaceTime enabled device.
- **Contacts** – Tapping on <u>contacts</u> will bring you to your contact list.

You can also browse through your iPad while on a call by simply pressing on the <u>home button</u>. To get back to the call screen itself mid-call, tap on the <u>green icon</u> at the top of your screen.

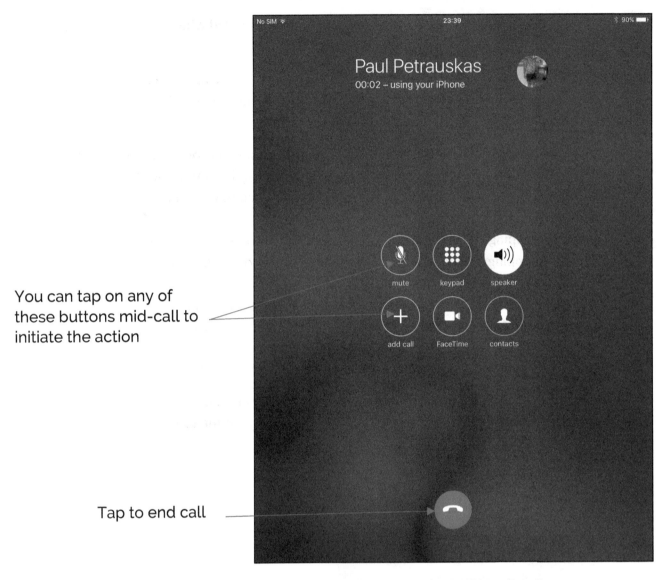

You can tap on any of
these buttons mid-call to
initiate the action

Tap to end call

Figure 19.5 – [On iPad] In-call options

Chapter 20 – Text Messaging

The iPad is a wonderful tool for text messaging. The ease in which to send and receive text messages is astounding, and the way texts are organized are even better. Using the Messages app, you can exchange text message with other Apple device users. If you also have an iPhone, you can use your iPad to exchange text messages with anyone.

Let's explore the Messages app on the iPad.

A Note about Messages on the iPad

With the Messages app, you can send and receive text messages with other people who use Apple devices. In order to use this service, you must be signed in with your Apple ID. If you also own and use an iPhone, you can send and receive text messages with anyone using your iPad.

iMessage

(See Figure 20.1)

iMessage is a special term used for text messages exchanged between people who use Apple devices. For instance, you may notice when texting certain people that your messages will appear as the color blue. Conversely, you will notice when texting other people that your messages will appear as the color green (Figure 20.2). When they appear blue, that means you are sending iMessages to each other. In other words, you both are using Apple devices such as an iPad, iPhone, or Mac to send your text messages. iMessages (blue) are different because they are not sent as cellular text messages. In other words, they do not count against your text message count if your cellular plan limits them. By default on the iPad, you will only be able to exchange text messages with people who use Apple devices, so your text conversations will be blue. If you also have an iPhone, you can enable Text Message Forwarding which will allow you to send and receive text messages with anyone on your iPad, including people who do not use Apple devices. The next section covers how to enable Text Message Forwarding on your iPhone. If you do not have an iPhone, you can skip that section.

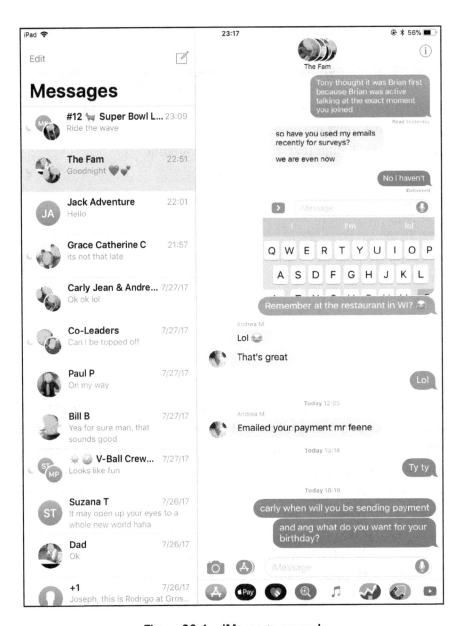

Figure 20.1 – iMessage example

Figure 20.2 – Regular non-Apple device text message

Setting up Text Message Forwarding on your iPhone (if Applicable)

In order to exchange regular text messages with people who do not use Apple devices on your iPad, you must setup Text Message Forwarding on your iPhone. To do so, follow these steps on your **iPhone**:

1. Open the Settings app.
2. Tap on Messages.
3. Tap on Text Message Forwarding.
4. Tap on the icon next your iPad or device that you want to be able to send and receive all text messages. (Figure 20.3)

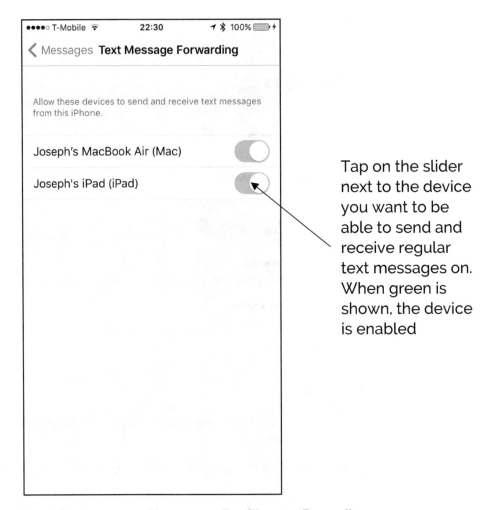

Tap on the slider next to the device you want to be able to send and receive regular text messages on. When green is shown, the device is enabled

Figure 20.3 – [On iPhone] Settings app -> Messages -> Text Message Forwarding

5. Tap the <u>back arrow</u> to go back a screen.
6. Tap on <u>Send & Receive</u>.
7. On this screen, make sure your Apple ID is shown and that you are signed in to both your iPhone and iPad using the same Apple ID. (<u>Figure 20.4</u>)
8. Make sure YOU CAN BE REACHED BY IMESSAGE AT: <u>Your Phone Number</u>, and <u>Your Apple ID Email</u> are both checked. You can tap on them to check them off. (<u>Figure 20.4</u>)
9. Make sure under START NEW CONVERSATIONS FROM: <u>Your Phone Number</u> is checked. If not, tap on it. (<u>Figure 20.4</u>)

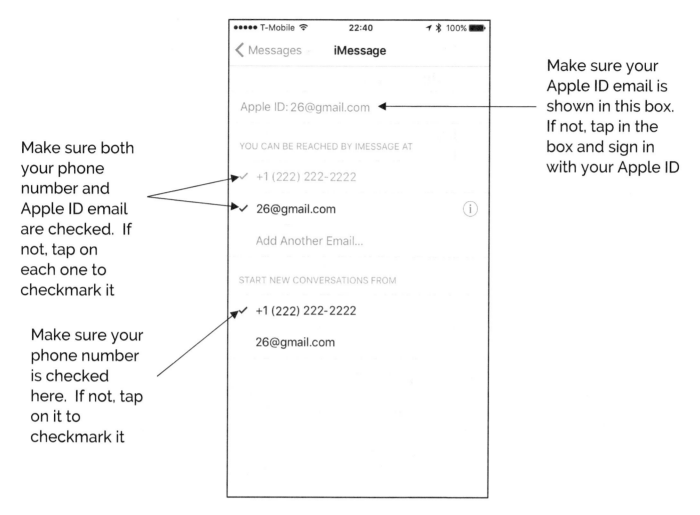

Make sure your Apple ID email is shown in this box. If not, tap in the box and sign in with your Apple ID

Make sure both your phone number and Apple ID email are checked. If not, tap on each one to checkmark it

Make sure your phone number is checked here. If not, tap on it to checkmark it

Figure 20.4 – [On iPhone] Settings app -> Messages -> Send & Receive

Now all text messages you receive on your iPhone you will also receive on your iPad. Furthermore, you will be able to exchange text messages with anyone on your iPad, including non-Apple device users

Using the Messages App (iPad)

Somewhere on your home screen will be the Messages app. Tap on it to open it up. This is where all of your text messaging will take place, and where all of your text messages will be stored.

Before we dive in, let me explain some texting jargon. When you exchange a text message with someone, it is generally called a conversation. In this book, we will call any text conversation with someone or a group of people a *thread*. (Figure 20.5)

Manage Messages
(Delete threads)

Start a New Text
Message (Called a
thread)

Unread messages
will have a blue icon.

These are all your *threads*
(text messages), sorted
by person or group.

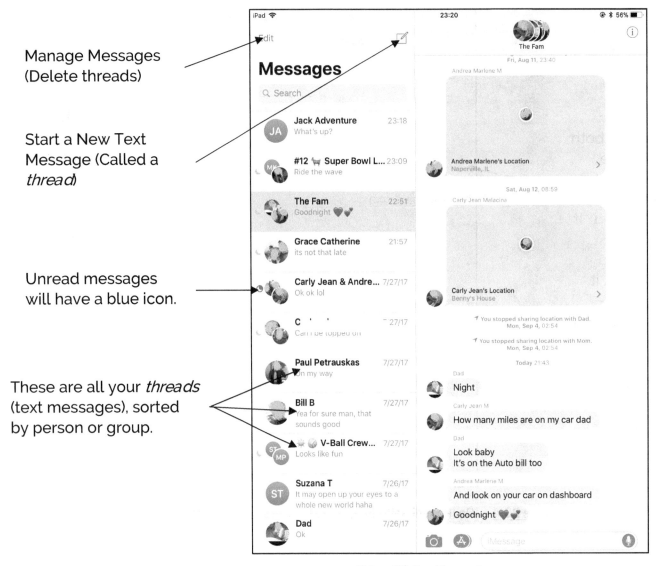

Figure 20.5 – Messages app

Sending a Text Message

To send your first text message on your iPad, first tap the Square and Pencil icon at the top of
your screen (Figure 20.5). This will start a new text thread. In the screen that appears, you can
type in the name of one your contacts and as you are typing, your iPad will make suggestions as
to whom you are trying to text. Once you see their name, you can tap on their *name* to
complete the entry (Figure 20.6). Names that appear in blue are people who are using an Apple
device and have iMessage enabled on their device. You can only send text messages to people
in blue unless you also have an iPhone.

Alternatively, you can use your keyboard to type in the exact phone number of someone you
want to text. (Only applicable if you have iPhone Text Message Forwarding enabled).

Now you can tap into the <u>Message bar</u> or <u>iMessage bar</u> near the middle of your screen. Now you can use your keypad to type in a message. Once that message is ready to be sent, tap the <u>up arrow icon</u> to send the message. (<u>Figure 20.6</u>)

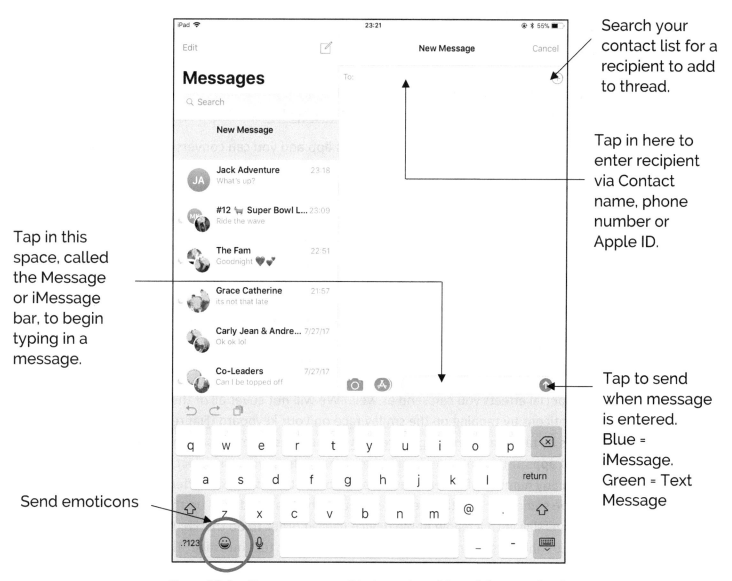

Search your contact list for a recipient to add to thread.

Tap in here to enter recipient via Contact name, phone number or Apple ID.

Tap in this space, called the Message or iMessage bar, to begin typing in a message.

Tap to send when message is entered. Blue = iMessage. Green = Text Message

Send emoticons

Figure 20.6 – Messages app -> Start new thread (pencil & square icon)

Receiving a Message

When you receive a message, it will appear in the Messages app. (See <u>Figure 20.5</u>). Unread messages will have a blue icon next to the thread. To view the message, open the thread by tapping on the *thread's name*. Your entire text message conversation will now be shown, including any new messages that you have received from this thread. You can scroll up and down through the history of the conversation using your finger.

Group Threads

Text messaging is not reserved for one-on-one conversations. You can create group threads where multiple people can exchange text messages. The process is exactly the same:

1. Tap the square and pencil icon at the top of your screen to create a new message
2. Enter in the name of a person you want to text, of the group. Tap on their *name* once it appears in suggestions.
3. Now type in another name of a person you also want to be a part of the group thread, and tap on their *name* in suggestions when it appears.
4. Add as many people to the thread as you wish.
5. When ready to start conversing, tap into the Message box and begin typing away.

This group thread will now be saved in your Messages app and you can converse in there any time you wish.

Special Effects

(See Figure 20.7)

There are a ton of special effects and add-ons you can do with the Messages app. For instance, if you want to text message a picture to someone, tap the camera icon to the left of the message bar. Now you can browse through your saved photos on the bottom by swiping left and right, and you can send one of those photos by tapping on it and then tapping the send icon. Alternatively, you can take a picture to send by swiping to the left and tapping on Camera. (Figure 20.7)

There are more special effects you can send as well. We will not cover all of them in here, but you can send emoticons by tapping on the smiley face on your keyboard (Figure 20.6). You can send more effects by tapping on the App Store icon to the left of the message bar and then tapping the bar of apps at the bottom of your screen. From here you can select from apps that allow you to send any number of different effects. Another type of special effect that is easy to send is to first type some text into the message bar and then tap and hold on the blue send icon (you must be conversing with a fellow Apple device user in order to do this i.e. iMessage). Now you can choose a bubble or screen effect to go along with your text message. These are fun to play around with.

Lastly, you can send voice messages to an iMessage thread by using the microphone to the right of the message bar. You will see this icon when you do not have any text entered. Take the time to play around with these features at your leisure to see all you can do with Messages. You will notice that some features can only be used when exchanging iMessages with another Apple user.

Name of
thread or
contact name

See thread
details
including
who is in
thread

Send audio
messages or
tap and hold
when text is
entered for
more
special
effects

Send
Photos

Open Full
Screen
Camera

Messages
App Store

View all
your Photos

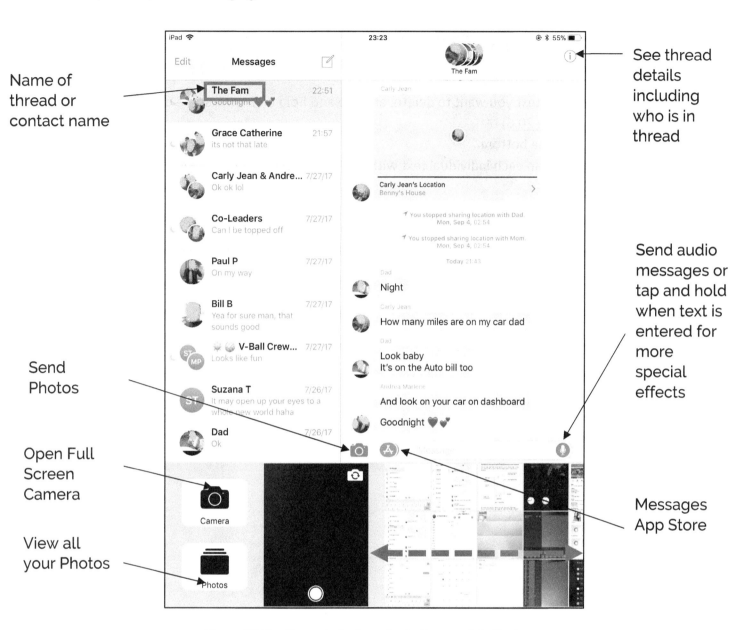

Figure 20.7 – Message features including special effects

Managing Messages

An important aspect of text messages is managing them. Let's take a look at a couple key management features that are prudent to know.

How to Delete Threads

You can delete an entire thread by following these steps:

1. Open the Messages app.
2. Tap Edit at the upper left
3. Tap each thread you want to delete, until they are check-marked.
4. Tap Delete at the lower left.

How to Delete Individual Texts

1. Open a thread in the <u>Messages app</u>.
2. Find the exact text you want to delete, and tap and hold on it until some options appear. (<u>Figure 20.8</u>)
3. Tap <u>More</u> at the bottom
4. Now you can tap each individual text within a thread to be check-marked that you want deleted.

5. Tap the <u>trash icon</u> at the lower left to delete the selected texts.

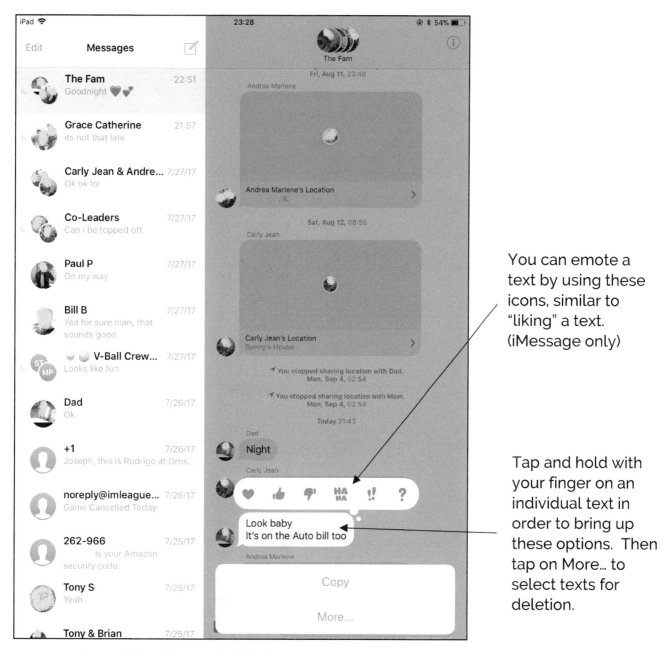

You can emote a text by using these icons, similar to "liking" a text. (iMessage only)

Tap and hold with your finger on an individual text in order to bring up these options. Then tap on More... to select texts for deletion.

Figure 20.8 – Managing individual text messages

Thread Details

Thread details offer a glimpse into certain information about a thread, such as who is in the thread, and any pictures exchanged in the thread. To access thread information, simply open a thread in Messages, and then tap the information icon at the upper right.

Chapter 21 – Tips & Tricks

Congratulations! You have made it through the bulk of this book, and you should now have a solid understanding of how to use your iPad. There are no more "basics" left to teach you, so I will leave you with a few tips and tricks that you may find helpful when using your iPad.

Backing up your Device

This is a MAJOR tip. You should always have a backup of your iPad's data available in case anything happens. If you destroy or lose your iPad, a backup will allow you to get all of your data back. There are two ways to back up your device.

Number One: Backup to iCloud (Recommended)

Backing up to iCloud is definitely the way to go. This is done automatically and continuously so you never have to worry about it. You will need an Apple ID to do so, and you must be signed in to iCloud with that Apple ID (covered in first few chapters of book). To turn iCloud Backup on, follow these steps:

1. Open Settings
2. Tap The Big Box at the top left with your name
3. Tap iCloud
4. Tap iCloud Backup
5. Now make sure iCloud Backup is on (enabled green). If it was off previously, tap on Back Up Now to begin the backup.

When your iPad is backing up to iCloud, you do not need to worry about setting up backup again, as your iPad will backup periodically, usually when you are sleeping and your iPad is charging. Please note your iPad must be connected to Wi-Fi in order to initiate its automatic backup. Should anything ever happen to your iPad, iCloud Backup will allow you to get your data back. iCloud saves your data securely in the "cloud".

Number Two: Backup to iTunes (More Advanced)

The second way to backup your device is to use iTunes. Doing this saves a backup on your computer that can quickly be restored. To learn how to do this, visit www.infinityguides.com and search for "iTunes" in Online Courses.

Automatically Update Apps

A very helpful and time-saving feature is the iPad's ability to update apps automatically when they need it. When this feature is on, you will not need to ever manually update apps through the App Store. Here are the steps:

1. Open <u>Settings</u>
2. Tap <u>iTunes & App Store</u>
3. Tap the button icon next to <u>Updates</u> under Automatic Downloads to enable it

Taking a Screenshot

A screenshot is a picture of exactly what your screen looks like, and can be very useful. For instance, say you received a text with instructions on how to bake a cake, and you want to share those instructions with a friend via text message. One way to accomplish this is take a screenshot of the original text and send the picture of that screenshot to your friend.

At any time you can take a screenshot of what you are viewing on your iPad by pressing and holding the <u>power button</u> and <u>home button</u> for one second, at the same time. You will know a screenshot was taken when you see your screen flash. Screenshots will save as pictures in your Photos app.

Restart your iPad

It is best practice to restart your iPad every now and then, the same way you would restart a computer to keep it fresh. I recommend restarting your iPad if it is running slow or not working properly. Doing so can fix the problem. As a good practice, I recommend restarting your iPad at least twice a month. To do so, simply turn the iPad off using the power button. Wait 60 seconds after the iPad is off, then turn it back on. You can also turn your iPad off from Settings in General.

Copy and Paste

Copying and pasting can be a very useful tool on your iPad. You can copy and paste pictures, text, links, and other items. For instance, to copy a website link in Safari, tap down the <u>rectangle with an arrow inside it</u> inside the top navigation bar. Then look at the second row of buttons and tap on the <u>Copy</u> button. You may have to scroll left and right to find it. Then to paste the link in a text message, double tap with your finger inside the <u>message bar</u> inside a thread in the Messages app, and then tap <u>Paste</u>.

To copy and paste text, find the text you want to copy and tap and hold on it. Now move the small marker dots left and right to select the exact portion of text you want to copy. Now tap the <u>Copy</u> box. You can paste the text by double tapping in a text box and then tapping <u>Paste</u>. (<u>Figure 21.1</u>)

Tap to copy

Tap and hold on text with your finger to bring up this selection tool

Drag the blue dots to expand over the selection of text you want to copy

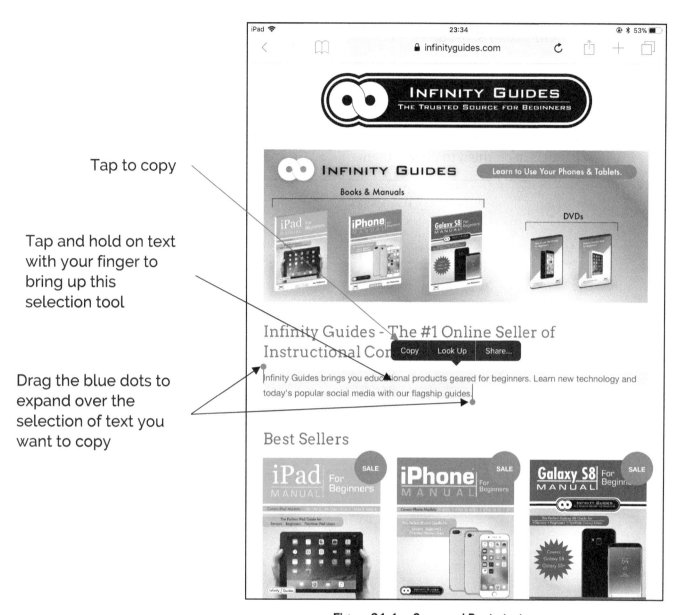

Figure 21.1 – Copy and Paste text

Multitasking & Gestures

The iPad has some unique features that can come in handy. One of these is multitasking, which lets you use two apps at the same time. To use this, you first must make sure Multitasking is enabled in Settings. (Figure 21.2)

How to Enable Multitasking & Gestures

1. Open the Settings app.
2. Tap General on the left.
3. Tap Multitasking & Dock inside the right pane.
4. Make sure the tabs next to Allow Multiple Apps and Gestures are highlighted green. If they are not, tap on them to enable them.

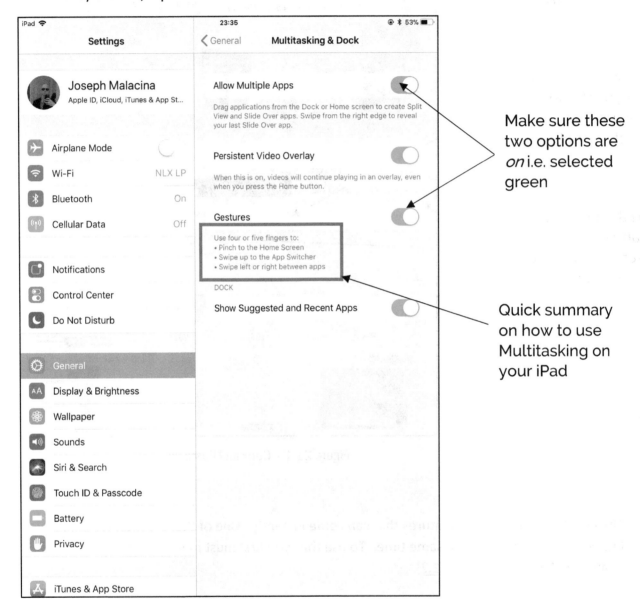

Figure 21.2 – Settings -> General -> Multitasking

While in Settings, you can read exactly how to use multiple apps and Gestures.

How to Use Two Apps at the Same Time

First, open an app you want to use such as the <u>Safari app</u>. Now, tap down at the bottom of your screen and swipe up slightly to bring up the Dock only (<u>Figure 21.3</u>).

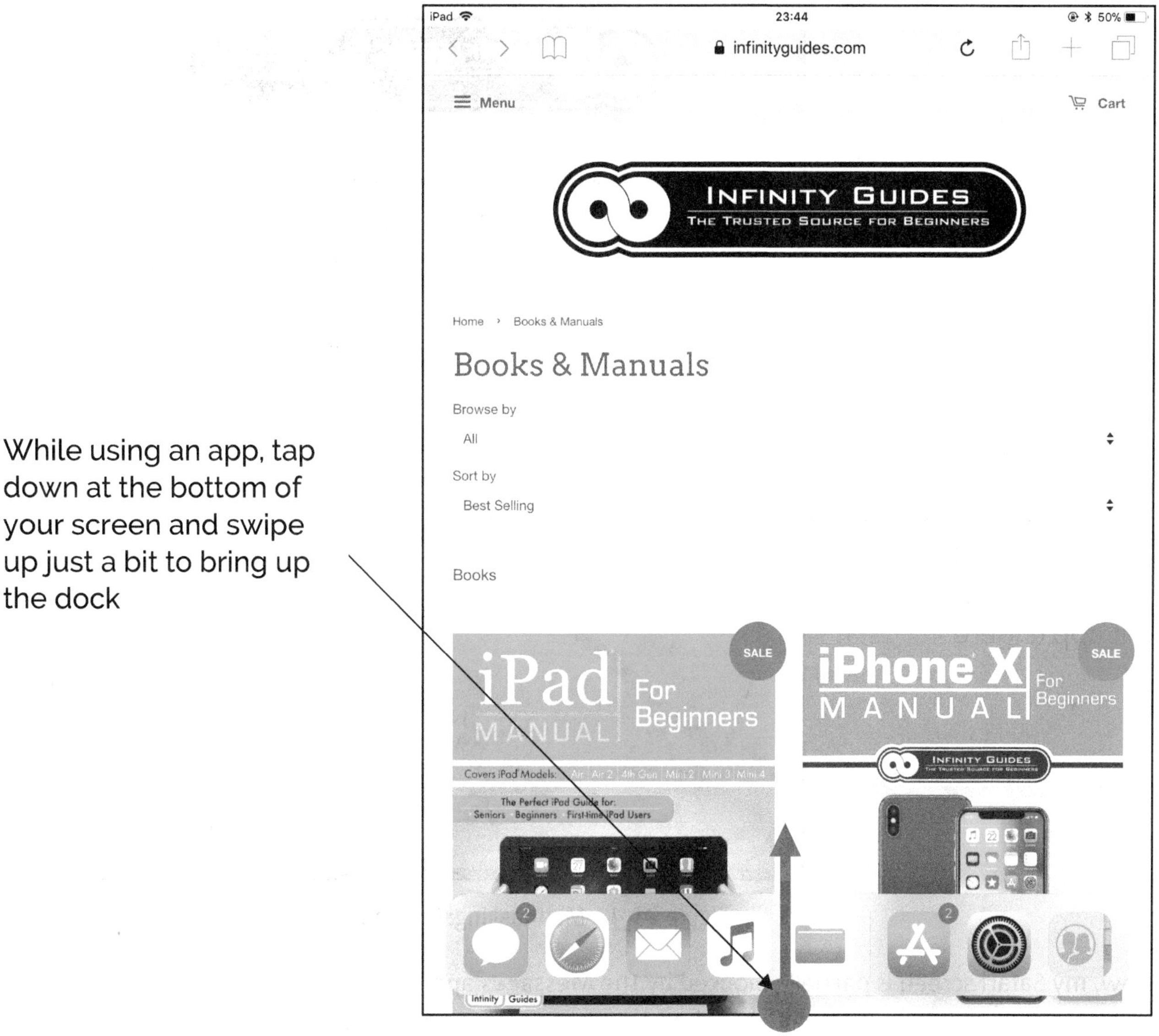

While using an app, tap down at the bottom of your screen and swipe up just a bit to bring up the dock

Figure 21.3 – Multitasking: Bring up the Dock

You swiped up too far if you brought up the full Control Center. Once the Dock is shown, tap on an app on the dock and hold for a second. Now drag that app up onto your Safari screen. You can now release the app on where you want it to appear alongside the other app. For instance, in this example I have the Safari app open. I brought up the Dock and then I grabbed and dragged the Messages app onto my Safari screen. If I let go and drop the Messages app icon somewhere on the right half of the Safari screen (<u>Figure 21.4</u>), the Messages app will be placed over my Safari screen.

Tap and hold on an app on your dock momentarily, then drag the app somewhere on your screen to place a small window of that app on top of your current view. If you drag the app all the way to the right edge, it will be places alongside your current app (Figure 21.6).

Figure 21.4 – Multitasking: Drag app from Dock

Now, my Safari screen is partially blocked by the Messages app, and I can use the Messages app by interacting with it. I can even move the Messages app to a different area by tapping on the horizontal line at the top of it and dragging it left and right. If I drag that horizontal line down, the Messages app will move to a split screen location with the Safari app (Figure 21.5). To completely get rid of the Messages app on top of the Safari screen, I can tap the edge or top of the Messages app screen and drag it all the way to the right and off the screen. This will completely hide the Messages app from view and can also be accomplished by simply pressing the home button. I can easily bring it back by simply tapping on the right edge of the screen and swiping left.

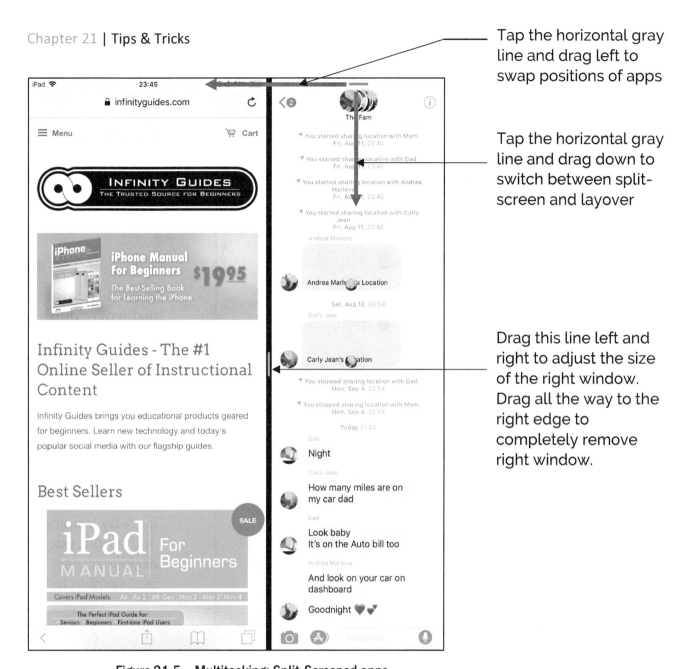

Tap the horizontal gray line and drag left to swap positions of apps

Tap the horizontal gray line and drag down to switch between split-screen and layover

Drag this line left and right to adjust the size of the right window. Drag all the way to the right edge to completely remove right window.

Figure 21.5 – Multitasking: Split-Screened apps

Now, when I first dragged the Messages app from the Dock to the Safari screen, I could have dropped the Messages app on the right edge of the Safari screen instead. This would have placed the Messages app in a split-screen position (Figure 21.5). With split-screen, no part of any app is blocked from view and I can adjust the size of the split-screen by tapping and dragging the vertical line at the center of the two screens. I can also tap on the horizontal line and drag it down to break up the split-screen and place the Messages atop the Safari screen instead. To completely remove the Messages app from the split-screen view, simply tap on the vertical line at the center of the split-screen and drag it to the right edge of the screen. When an app is on top of another app (not split-screen), you can drag its left edge all the way to the right to hide it (Figure 21.6). To bring it back, tap on the right edge of your screen and swipe left.

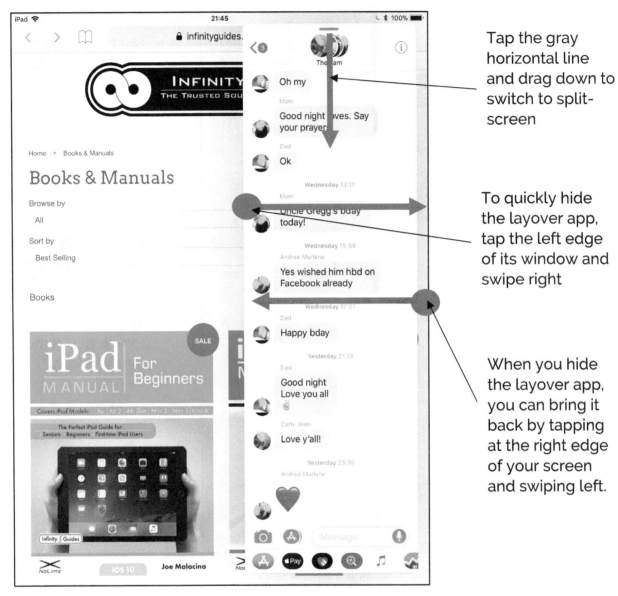

Tap the gray horizontal line and drag down to switch to split-screen

To quickly hide the layover app, tap the left edge of its window and swipe right

When you hide the layover app, you can bring it back by tapping at the right edge of your screen and swiping left.

Figure 21.6 – Multitasking: Messages app layover Safari app

I admit it is difficult to express how to use these multitasking features in text and it can even be confusing trying it yourself. Regardless, the multitasking feature can be incredibly helpful when you want to see two apps at the exact same time, and I recommend that the first time you try to use the split-screen function that you have this book handy and open to this section for guidance. I would also recommend using split-screen in landscape orientation for best results.

Gestures

There are several gestures you can use to quickly perform some actions. You can quickly switch between recently used apps (Figure 21.7), "pinch" to return to the home screen (Figure 21.8), and bring up recently used apps, also called the "app switcher" (Figure 21.9).

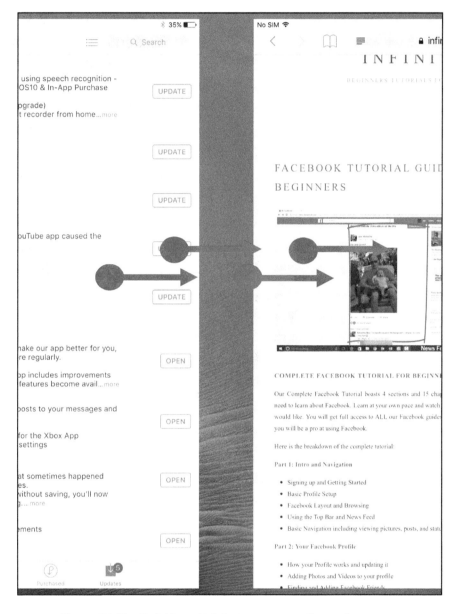

To quickly switch between recently used apps, tap down on your screen while using an app (not the home screen) with four fingers relatively close to each other. Then swipe all four of your fingers either left or right.

Figure 21.7 – Quickly switch between recently used apps

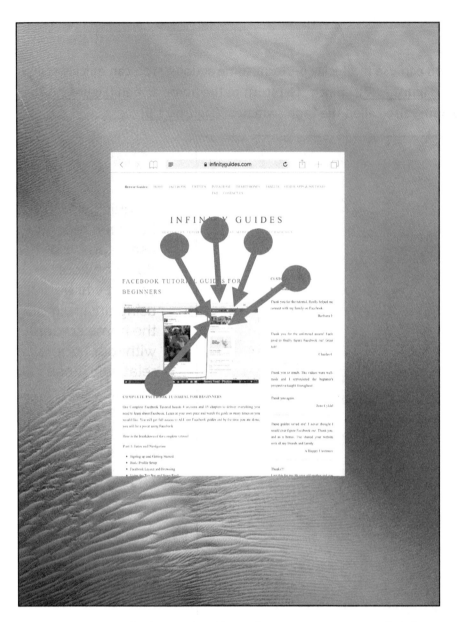

The *pinch to home screen* gesture is just a fancy way to get back to your home screen while using an app. To do this, while using an app (not the home screen), place all five of your fingers including your thumb on the middle of the screen. Then bring all your fingers together as if making a fist. This will bring you to your home screen.

Figure 21.8 – Pinch to Home Screen gesture

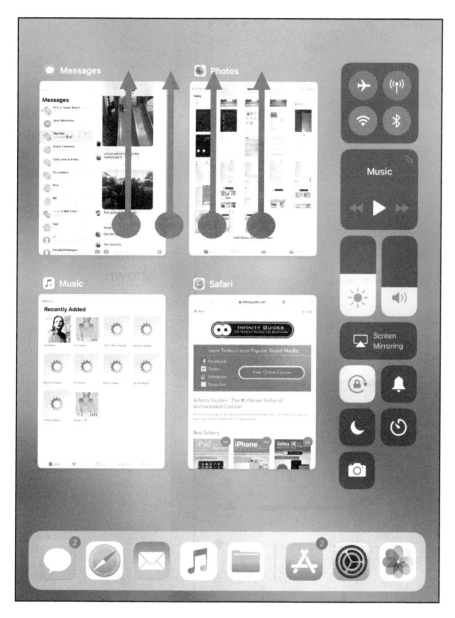

You've seen this screen before in *Chapter 16*. You can quickly get to this screen while using an app by placing four fingers on the screen, and then dragging all four fingers upwards towards the top of the screen. You can use this gesture on your home screen as well.

Figure 21.9 – Shortcut to Control Center

Splitting the Keyboard

This little known trick allows you to split the keyboard on your iPad in half so you can type with your thumbs. This tip is not for everyone, but here is how to do it just in case. (Figures 21.10 & 21.11).

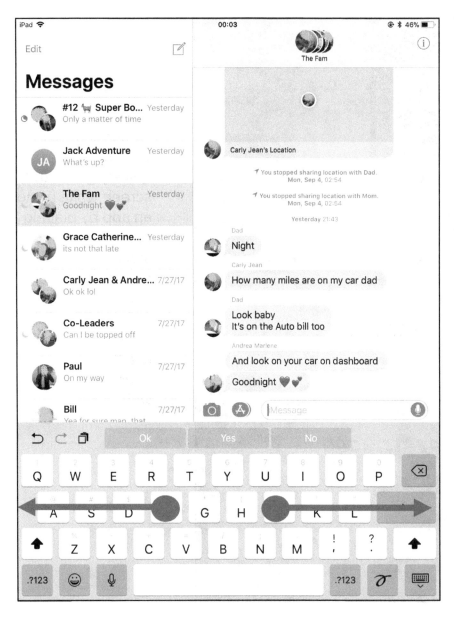

While your keyboard is up, such as when using the Messages app, tap with two fingers on the keyboard, then drag them apart horizontally as shown.

<u>Figure 21.10</u> – Splitting the keyboard

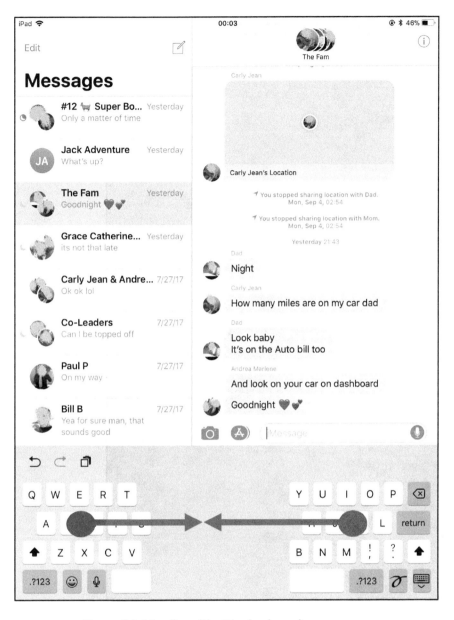

Your keyboard will now be split in half! This is a great feature for people who like to type with their thumbs.

To bring the keyboard back together, tap down on each side of the keyboard with a finger, then drag your fingers back together horizontally as shown.

Figure 21.11 – Reuniting the keyboard

The Lock Screen

Your lock screen is the screen you see when you awaken your iPad from the sleep state (Figure 21.12). If you setup a passcode or Touch ID, you cannot leave the lock screen until you have passed through your security.

Figure 21.12 – The Lock Screen

Your lock screen is available for anyone to see, even people who do not know your password. There are steps you can take to hide content that you do not want to appear on your lock screen. Here's how:

Restricting Access to your Lock Screen (Figure 21.13)

1. Open the Settings app.
2. Tap Touch ID & Passcode
3. Enter your passcode
4. Adjust the settings under ALLOW ACCESS WHEN LOCKED

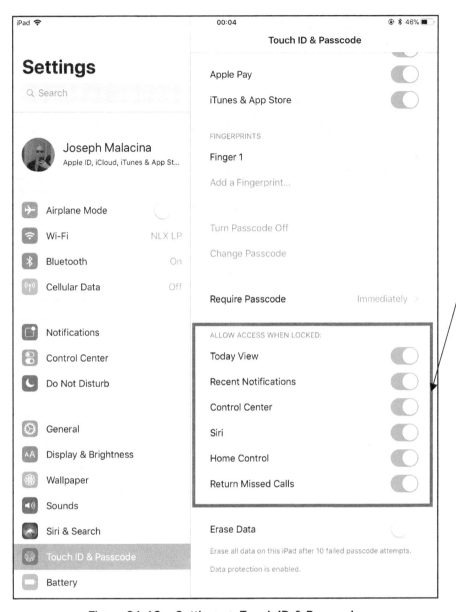

Here you can choose what you want to allow access to on your lock screen while your iPad is locked.

Figure 21.13 – Settings -> Touch ID & Passcode

Another aspect you may want to control is text message previews. By default, when you receive a text message on your iPad you can read a preview of it right on your lock screen. This also means that when you leave your iPad at home and you get a text message, anyone at your home can read that preview as well. You can protect against this by hiding text message content.

Hiding Text Message Preview Content (Figure 21.14)

1. Open Settings
2. Tap Notifications on the left
3. Find the Messages app on the right and tap on it

4. Tap on <u>Show Previews</u> under MESSAGE OPTIONS
5. Tap on <u>When Unlocked</u> or <u>Never</u>

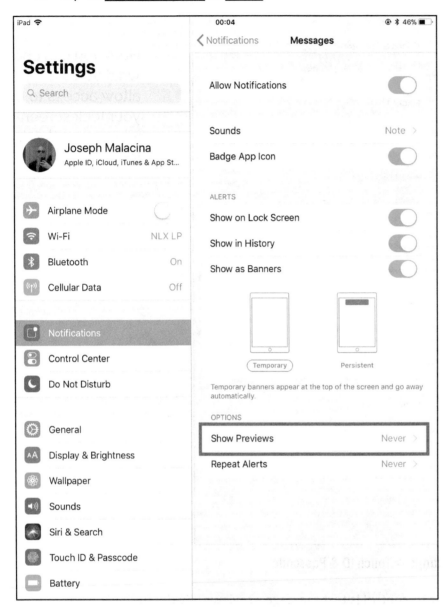

<u>Figure 21.14</u> – Settings -> Notifications -> Messages

Accessibility Options

Your iPad comes with several accessibility options to help you use your iPad if you have certain requirements. These include options for vision and hearing impairments. To browse through the different accessibility options, open the <u>Settings</u> app, tap <u>General,</u> and then tap <u>Accessibility</u>.

From here you can browse through different accessibility features. One common feature is <u>Larger Text</u>, which makes all text appear larger on your iPad. Another useful feature is <u>AssistiveTouch</u>, which helps if you have difficulty touching down on the screen.

The News App

The News App is a native app to the iPad that brings several outlets of news all together into one app. When you first open the <u>News</u> app you will be brought through an introduction and asked how you would like to use the app. Follow the instructions on the screen to setup news alerts and notifications. If you want to get your news from specific sources, you can always download that news organization's app. You can also see snippets from the News app on the Today screen, which can be accessed from your home screen by swiping all the way to the right.

Shutting an App Down

At some point you are bound to come across an app that is not working properly. It may be frozen or performing very slowly. Sometimes, the only way to fix this is to completely close the app down and then launch it back up again. Whenever you press the home button to exit an app, that app does not completely close down. There is nothing wrong with this, as this is how the iPad and its iOS are designed to function. However, should you ever find the need to completely close down an app, here is how:

1. Open the Control Center by pressing the <u>home button</u> twice
2. All of your recently used apps will show on the left hand side. You can swipe left and right to view them all.
3. To close one or more of these completely, tap on the screen of an app in the Control Center and swipe it up and off the screen. That app will completely shut down. Do this for each app that you want to shut down.
4. Press the <u>home button</u> to return to your home screen when done.

Siri Easter Eggs

As mentioned in the Siri chapter, you can say anything you want to Siri. Anything. You may even discover some Easter eggs while speaking to her. Try these: (Press and hold the <u>home button</u> to begin).

- How much wood could a woodchuck chuck if a woodchuck could chuck wood?
- What do you look like?
- I love you.
- I love my iPad.
- I am tired.

Try some more yourself.

Chapter 22 – More Resources

This guide has covered all the beginner aspects of the iPad. We have also covered many intermediate and advanced aspects, but there is still plenty more you can learn. Most notably, there is a lot to learn about specific apps. You can also learn more about Apple Music, iTunes, and social media on your iPad.

Infinity Guides is an excellent resource for beginners, and on the Infinity Guides website you can find books, manuals, DVDs, and online courses made for beginners. The online courses can be especially helpful as most of them are about 30 minutes long and teach you through video instruction. I have created a short list of things you can learn with Infinity Guides for your reference.

www.infinityguides.com **Content:**

- Facebook for Beginners
- Twitter for Beginners
- Apple Music for Beginners
- Smartphones & Tablets for Beginners
- Instagram App for Beginners
- Snapchat App for Beginners
- iTunes for Beginners
- Mac Computer for Beginners
- Making your Computer Fast Again Tutorial

CONCLUSION

Thank you for taking the time to read this text. It is my hope that you feel much better about using your iPad. I am confident that if you took the time to read this entire book, then you will have no problem using every aspect of your iPad with ease. Continue to use this book as a reference when you need it. The table of contents can quickly lead you to your answer, and the appendixes can be especially helpful as well and I hope you use them.

Be sure to explore various apps in the App Store so you can truly discover all you can do with your iPad. I urge you to use the Appendix in this text to see some great apps and explore all that you can do with your iPad.

I welcome your thoughts and feedback on this book; please come visit my Facebook page online at www.facebook.com/joemal. You can also tweet me @JoeMalacina on Twitter. I am often online answering questions from people who have read this text, and helping people with complex iPad issues. As a last piece of advice, please remember your Apple ID, Apple ID password, and lock screen passcode. Forgetting even one of these, especially your lock screen passcode, can be a real headache.

Enjoy using your iPad.

APPENDIX A – Additional Recommended Apps

- **560 The Answer** – A talk radio station's app that allows you to listen live and listen later
- **WLS 890** – Another talk radio station's app that allows you to listen live and listen later.
- **AirBnB** – Book lodging from regular people who put their vacation homes, rentals, condos, and dwellings up for short-term lodging.
- **Amazon** – Shop with Amazon's app
- **Bible** – App for reading the Bible
- **BigOven** – Cooking & Recipes
- **Clash of Clans** – World-wide simple strategy game
- **Dropbox** – Online File Storage and Sharing
- **EverNote** – An app for taking and saving notes securely
- **Expedia** – An app to lookup flight, hotel, and car prices
- **Facebook** – Social Media
- **Facebook Messenger** – The Messenger app for Facebook
- **Fandango** – Lookup movie show times, reviews, information, and local theaters
- **Flixster** – Movie ticketing app
- **Fox News** – The official app for the Fox News Channel
- **Fox Sports** – The official app for the Fox Sports channel
- **Google Chrome** – An alternative web browser
- **Google Maps** – Navigation app for GPS, directions, and places
- **Groupon** – An app that allows you to reserve coupons and deals for products and services
- **HD Wallpapers** – App to download more wallpapers for your iPad
- **Kindle** – Download, rent, and purchase books through the Kindle service
- **McDonalds** – McDonalds' official app. Sometimes they even have good coupons and deals on there.
- **Microsoft Office** – Apps for Word, Excel, and other Microsoft Office products.
- **Netflix** – Subscription app that allows you to watch various TV, movies, documentaries, and original content straight from your iPad.
- **Nook** – An app for reading and downloading eBooks.
- **OpenTable** – App for finding and booking reservations at restaurants.
- **Radio.com** – Listen to various radio stations around the globe.
- **Shazam** – Allows to identify any song that is playing on the radio or other format.

- **Shopular** – App for finding discounts at retail stores.
- **Skype** – Video chat with your friends and family.
- **Speedtest** – Run networking tests on your connection.
- **Target** – Target's official app
- **The Weather Channel** – My favorite weather app.
- **Twitter** – Social media app
- **Uber** – App that lets you hail rideshares and taxis.
- **Venmo** – App that lets you pay your friends and family instantly.
- **VLC** – Video Player
- **Wall Street Journal** – News app
- **WebMD** – Health app
- **WhatsApp** – Worldwide Calling and Texting
- **Yelp** – App that shows reviews of places.
- **Your Bank's App** – Can come in handy for checking balances, depositing checks, and making transfers.
- **Your Cable TV App** – You may be able to access your TV stations from your iPad.
- **Your TV's App** – Some TV's allow you to download a remote app on your iPad.

Many people ask how I came up with this list of recommended apps. Well, the answer is I have used and tested every single one of the apps I have recommended. Many of the apps I use every single day. My criteria for recommending an app to you is simple: it must be easy to use, free of bugs and glitches, have a strong user base, not be "spammy", and most importantly, the app must have an incredibly useful function. I also included a couple of hidden gems in the list that may not be widely known, but nevertheless are great apps. If you have a recommendation you would like to share with me I would love to hear it! Send me a message on Twitter @JoeMalacina.

APPENDIX B – Siri Examples

These are all examples of things you can say to Siri.

- What time is it?
- Show me restaurants around here
- Email *Contact Name*
- Send a text to *contact name*
- Open *app name*
- Check my e-mail
- Check my messages
- Do I have any appointments today?
- Check my battery life
- Tell *contact name* I am on my way in a text
- What movies are playing?
- Play *song name*
- Play *playlist name*
- Play *artist name*
- Remind me to wash the dishes tonight
- Email *contact name* about the trip
- Wake me up at 7 AM tomorrow.
- Note that I need to get my dog a new bone
- Turn on Airplane mode
- Decrease/Increase my brightness
- What is Apple's stock price?
- What is the address of *contact name*?
- Change my wallpaper
- Play a random song
- Listen to the news
- Where can I get a burger around here?

APPENDIX C – List of Common Functions

- **Add New Email to iPad:** Settings -> Accounts & Passwords -> Add Account
- **Bookmark a Web Page:** Safari App -> Square and Rectangle Icon -> Add Bookmark
- **Change Language:** Settings -> General -> Language & Region
- **Change Lock Screen Password or Fingerprints:** Settings -> Touch ID & Passcode
- **Change Ringtones:** Settings -> Sounds (& Haptics)
- **Change Wallpaper:** Settings -> Wallpapers
- **Check for iOS Updates:** Settings -> General -> Software Update
- **Completely restore iPad to Factory Default (WARNING: this will delete all of your iPad data and bring it back to right out of the box status. DO NOT DO THIS unless you know what you are doing.):** Settings -> General -> Reset -> Erase all Content and Settings - > enter passwords
- **Connect to a Bluetooth Device:** Settings -> Bluetooth -> My Devices
- **Connect to a Wi-Fi Network:** Settings -> Wi-Fi -> Tap on Network -> Enter password -> Tap Join
- **Create New Text:** Messages -> Square and Pencil Icon
- **Join Apple Music:** Settings -> Music -> Join Apple Music
- **Set Default Search Engine:** Settings -> Safari -> Search Engine
- **Sign in to your Apple ID:** Settings -> iTunes & AppStore -> Apple ID
- **Switch to Military Time:** Settings -> General -> Date & Time -> 24 Hour Time
- **Turn Location Services On/Off:** Settings -> Privacy -> Location Services
- **Turn on Do Not Disturb:** Control Panel -> Moon Icon
- **Turn on iCloud Photo Library:** Settings -> Photos -> iCloud Photo Library
- **Turn Siri On/Off:** Settings -> Siri & Search
- **Create a new E-mail:** Mail app -> Square and Pencil Icon
- **Forgot Your Passcode?:** Go to https://support.apple.com/en-us/HT204306
- **Forgot your Apple ID password?** Go to https://support.apple.com/en-us/HT201487

Notes

Check out more beginner's guides and manuals at:

www.infinityguides.com